可持续性设计（影印版）
Designing for Sustainability

Tim Frick 著

Beijing · Boston · Farnham · Sebastopol · Tokyo

O'Reilly Media, Inc.授权东南大学出版社出版

南京 东南大学出版社

图书在版编目(CIP)数据

可持续性设计:英文/(美)提姆·弗里克(Tim Frick)著. —影印本. —南京:东南大学出版社,2018.2
书名原文:Designing for Sustainability
ISBN 978-7-5641-7468-2

Ⅰ.①可… Ⅱ.①提… Ⅲ.①网页-程序设计-英文 Ⅳ.①TP393.092

中国版本图书馆 CIP 数据核字(2017)270637 号
图字:10-2017-344 号

© 2016 by O'Reilly Media, Inc.

Reprint of the English Edition, jointly published by O'Reilly Media, Inc. and Southeast University Press, 2018. Authorized reprint of the original English edition, 2017 O'Reilly Media, Inc., the owner of all rights to publish and sell the same.

All rights reserved including the rights of reproduction in whole or in part in any form.

英文原版由 O'Reilly Media, Inc.出版 2016。

英文影印版由东南大学出版社出版 2018。此影印版的出版和销售得到出版权和销售权的所有者 —— O'Reilly Media, Inc.的许可。

版权所有,未得书面许可,本书的任何部分和全部不得以任何形式重制。

可持续性设计(影印版)

出版发行:东南大学出版社
地　　址:南京四牌楼 2 号　　邮编:210096
出 版 人:江建中
网　　址:http://www.seupress.com
电子邮件:press@seupress.com
印　　刷:常州市武进第三印刷有限公司
开　　本:787 毫米×980 毫米　　16 开本
印　　张:23.25
字　　数:402 千字
版　　次:2018 年 2 月第 1 版
印　　次:2018 年 2 月第 1 次印刷
书　　号:ISBN 978-7-5641-7468-2
定　　价:94.00 元

本社图书若有印装质量问题,请直接与营销部联系。电话(传真):025-83791830

[*contents*]

Preface ix

Chapter 1 **Sustainability and the Internet**......................... 1
What You Will Learn in This Chapter.................. 1
A Greener Internet 1
Sustainability Defined................................. 4
Sustainability in Business 9
Sustainability and the Internet 28
Virtual LCAs 41
Conclusion..48
Action Items...48

Chapter 2 **A Sustainable Web Design Primer** 49
What You Will Learn in This Chapter................49
Sustainable Web Design.............................49
Conclusion..65
Action Items...65

Chapter 3 **Sustainable Components** 67
What You Will Learn in This Chapter................67
A Greener Apple......................................67
Other Sustainable Components 82
Potential Barriers and Workarounds 98
Conclusion...105
Action Items..105

Chapter 4	**Content Strategy** 107
	What You Will Learn in This Chapter 107
	The Content Conundrum 107
	Potential Barriers and Workarounds 147
	Conclusion..148
	Action Items..149
Chapter 5	**Design and UX** 151
	What You Will Learn in This Chapter.................151
	Users Versus Life Cycles..............................151
	Visual Design...167
	Web Standards..183
	Potential Barriers to Sustainable UX 188
	Conclusion..188
	Action Items..189
Chapter 6	**Performance Optimization** 191
	What You Will Learn in This Chapter................191
	Performance Counts..................................191
	A Balancing Act 194
	Speed Is Just One Metric............................ 200
	Performance Rules 210
	Speed, Reliability, and Version Control...............214
	Workflow Tips 221
	Accessibility and Sustainability 225
	Potential Barriers and Workarounds 227
	Conclusion... 228
	Action Items... 228
Chapter 7	**Digital Carbon Footprints** 229
	What You Will Learn in This Chapter................229
	Estimating a Carbon Footprint......................229
	Proposing a Framework 234
	Ecograder: A Case Study.............................241

	Ecograder Benchmarking	257
	Conclusion	261
	Action Items	261
Chapter 8	**A Future-Friendly Internet**	**263**
	What You Will Learn in This Chapter	263
	Future-Friendly Web	264
	Interviews: Industry Leaders Predict	279
	Final Word	284
	Appendix A: Figure Attributions and Links	285
	Acknowledgments	291
	Index	317
	About the Author	333

This book is dedicated to the team at Climate Ride and to the global B Corp community. You inspire me every day.

[*Preface*]

| Building a Cleaner, Greener Internet

Goliath and Frank: A Holiday Tale

As I write this in upper Michigan during the last days of 2015, the year is poised to take the title of hottest on record,[1] broken from the previous year in 2014. I recently rode my bicycle on snow-free roads, wearing a light jacket just a few miles from what has been named by The Weather Channel as the third snowiest city in the United States.[2]

FIGURE P-1.
A balmy December day near the third snowiest place in the United States

---

1 Jonah Bromwich, "A Fitting End for the Hottest Year on Record", New York Times, December 23, 2015. (*http://www.nytimes.com/2015/12/24/science/climate-change-record-warm-year.html?_r=1*)
2 Michael H. Babcock, "Hancock Named '3rd Snowiest City'", *The Daily Mining Gazette*, December 20, 2010. (*http://www.mininggazette.com/page/content.detail/id/518181/Hancock-named--3rd-snowiest-city-.html?nav=5006*)

In fact, many places in the eastern United States reported the warmest Christmas Eve on record, with Fahrenheit temperatures ranging from 20° to 30° higher than average.[3] New York City, Philadelphia, and other eastern seaboard cities reported temperatures in the low to mid-70s. Burlington, Vermont reported its highest December temperature ever at 68°. In total, more than 6,000 temperature records were broken in the United States alone during December 2015.[4]

Shortly thereafter, a single storm system—aptly named Winter Storm *Goliath* by The Weather Channel (but called *Frank* by weather agencies in the United Kingdom and Ireland[5])—encompassed more than half the United States. With it came a string of at least 55 tornadoes over a seven-day period that destroyed numerous homes and killed 20 people,[6] making it the deadliest December for tornadoes in 62 years.[7] Record-breaking blizzards in western Texas and New Mexico brought heavy snowfall and drifting of epic proportions, causing New Mexico to declare a state of emergency. One New Mexico couple was stranded in their car for nearly 20 hours under 12 feet of snow. Texas and New Mexico dairy farmers lost an estimated 30,000 cattle during the storm.[8] Goliath also brought the worst flooding to hit the Mississippi River since the Great Flood of 1993, in some places even surpassing those flood levels.[9] It took well into January for water levels in the mid-South and lower Mississippi valleys to subside. Missouri declared a state of emergency, as well. In addition, Goliath brought snow and hazardous

---

3 Brett Rathbun, "Warm Christmas Eve Shatters Records Across Eastern US", AccuWeather.com, December 26, 2015. (http://www.accuweather.com/en/weather-news/warmth-record-high-temperatures-northeast-southeast-christmas-2015/54388777)

4 Chris Dolce, "6 Incredible Facts About December's Warmth", The Weather Channel, December 22, 2015. (http://www.weather.com/news/weather/news/weird-facts-december-2015-warmth)

5 Wikipedia, "2015–16 UK and Ireland Windstorm Season". (https://en.wikipedia.org/wiki/2015%E2%80%9316_UK_and_Ireland_windstorm_season)

6 The Weather Channel, "Tornadoes and Flooding Rain Hit the South, Midwest Christmas Week 2015", December 28, 2015. (http://www.weather.com/storms/tornado/news/storms-tornadoes-christmas-week-december-21-28-2015)

7 The Weather Channel, "Tornadoes: Deadliest December in 62 Years", December 28, 2015. (http://www.weather.com/news/weather/video/tornadoesdeadliest-december-in-62-years)

8 Ada Carr, "Dairy Cow Death Toll to Surpass 30,000 in Texas, New Mexico Due to Winter Storm Goliath", The Weather Channel, January 1, 2016. (http://www.weather.com/news/news/dairy-cows-winter-storm-goliath-texas-new-mexico)

9 Jon Erdman, "Historic Winter Flood Along Mississippi River Sets Record in Cape Girardeau", The Weather Channel, January 6, 2016. (http://www.weather.com/news/news/mississippi-river-flooding-december-2015)

travel conditions throughout the Midwest and Northeast United States, canceling more than 2,800 flights with another 4,800 delayed at airports across the country, leaving hundreds of thousands without power.[10]

Not surprisingly, the storm left a swath of death and destruction in its wake, with at least 52 people killed and an as yet indeterminate amount of damage to property. Although The Weather Channel called it the deadliest storm of 2015,[11] the National Centers for Environmental Information (NCEI) noted that it was just one of *ten* storms in 2015 that exceeded the $1 billion mark in losses.[12]

The storm's path then continued into Europe, and its name was changed to Frank as it gained momentum from warmer waters of the Gulf Stream. By the time Frank hit Iceland, it had atmospheric pressure similar to some of the strongest hurricanes ever recorded, putting it into the top five of all Northern Atlantic storms.[13] In Northern England, Wales, Scotland, and Ireland, high winds, waves, and rain pounded an area already dealing with historic flooding. Over the North Sea, the storm caused an oil barge to break loose and drift out of control in the violent waves, costing one life and forcing BP to evacuate its Valhall oil field.[14]

When Goliath/Frank hit the North Pole, where average December temperatures hover around −15° F to −20° F below zero, the warm air from the Atlantic tropics pushed temperatures as high as 32+° F *above* zero, a temperature fluctuation of 50+° F higher than normal, something that

---

10 The Associated Press, "Latest: More Than 2,800 Flights Canceled Amid Winter Storm", December 29, 2015. (*http://bigstory.ap.org/article/19191dfd426042d5979cafa87e188a40/latest-weather-forces-i-40-closure-new-mexico-texas*)

11 Andrew MacFarlane, "Goliath: The Deadliest U.S. Storm System of 2015", The Weather Channel, December 31, 2015. (*http://www.weather.com/news/news/goliath-deadliest-storm-of-2015*)

12 National Centers for Environmental Information, "Billion-Dollar Weather and Climate Disasters: Table of Events". (*http://www.ncdc.noaa.gov/billions/events*)

13 Andrew Freedman, "Historic Storm Set to Slam Iceland, Northern UK with Hurricane-Force Winds", Mashable, December 28, 2015. (*http://mashable.com/2015/12/28/freak-atlantic-storm-uk-frank/#4CbtAB7_haqJ*)

14 Don Melvin, "One Killed, Oil Rigs Evacuated, Barge Drifts Loose in Violent North Sea", CNN, December 31, 2015. (*http://www.cnn.com/2015/12/31/europe/bp-evacuation-north-sea-oil-field*)

has been seen only three times since 1948.[15] Let that soak in for a second: the North Pole was above freezing at a time when 100% of its days are bathed in darkness.

Three weeks later, another major storm blanketed the mid-Atlantic and Northeast with up to 42 inches of snow, left 250,000 people without power, closed down New York City and Washington, DC, and left up to 48 people dead.[16] The estimated economic impact of that storm was $850 million.

Think these stories sound like something out of a Hollywood movie like *The Day After Tomorrow*? Unfortunately, this is not the stuff of fiction. Although some scientists warn of exploiting extreme weather events to promote climate change,[17] the majority agree that as global temperatures rise over time,[18] so does the likelihood of more intense storms, something the world clearly saw with Goliath/Frank.

So what, pray tell, does any of this have to do with becoming a better designer? (That *is* why you picked up this book, right?)

---

15 See meteorologist Bob Henson's December 28, 2015 tweet. (https://twitter.com/bhensonweather/status/681685436264132608)

16 Sean Breslin, "Winter Storm Jonas: At Least 48 Dead; Roof Collapses Reported; D.C. Remains Shut Down", The Weather Channel, January 26, 2016. (https://weather.com/storms/winter/news/winter-storm-jonas-impacts-news)

17 John Michael Wallace, "Confronting the Exploitation of Extreme Weather Events in Global Warming Reporting", *The Washington Post*, February 28, 2014. (https://www.washingtonpost.com/news/capital-weather-gang/wp/2014/02/28/confronting-the-exploitation-of-extreme-weather-events-in-global-warming-reporting)

18 Union of Concerned Scientists, "Is Global Warming Linked to Severe Weather?". (http://www.ucsusa.org/global_warming/science_and_impacts/impacts/global-warming-rain-snow-tornadoes.html#.VogKaxqAOko)

Weather Versus Climate

Weather is something that happens outside your window on any given day. It is not climate, which is measured over an extended period of time. A single extreme example, such as the one I just shared, does not a point prove, but this does consider the chart provided in Figure P-2.

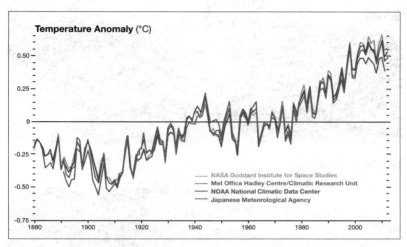

FIGURE P-2.
Temperature trends from four international science institutions, 1880 to present

According to NASA's global climate change website, 97% of scientists agree that "climate-warming trends over the past century are very likely due to human activities." Global warming trends have been reported as far back as the 1930s, with significant discoveries made every decade since, repeatedly proving that greenhouse gases like carbon dioxide, methane, nitrous oxide, carbon monoxide, and others released into the atmosphere through the burning and production of fossil fuels, deforestation, industrial agriculture, and other man-made practices have an environmental impact. Compounding this, as global temperatures rise, environments warm up. Some, like the Arctic tundra, have been storing carbon for thousands of years. As the ice in these environments thaws, that carbon is also released, exacerbating an already alarming situation.

FIGURE P-3.
Each year, hundreds of everyday people hike and bike across the United States to raise money for their favorite environmental nonprofit on Climate Ride

CONNECTING THE DOTS

How do we connect the dots between extreme weather stories and designers building mobile apps? They don't seem related, right? Well, everything we do has some sort of impact. Yes, something as innocuous as designing a web page, buying a coffee, eating dinner, or even breathing produces waste of some sort. Collectively, when applied to the 7.4+ billion people on the planet, that impact really adds up.[19]

One of the many challenges associated with the climate crisis is connecting the dots between extreme weather events like those noted earlier and personal or organizational behavior. When you bone-up on climate science basics and realize the depth of the situation we're in, it can feel overwhelming and hopeless. One wonders how a single person or small business could possibly have an impact on something as monumental as a melting glacier, rising seas, or years-long drought.

When you compound that with a political and economic climate that can often fuel personal apathy, it is not surprising that many don't feel compelled to act. Here are two examples:

- Senator James Inhofe, 2015 chair of the US Senate Committee on Environment and Public Works—mistaking weather for climate (see the sidebar on page xiii)—threw a snowball to the sitting Senate President on the Senate floor in an effort to disprove global warming because it was unseasonably cold outside.[20]
- An investigation in 2015 unearthed evidence that Exxon (now ExxonMobil) was aware of climate change's impact as early as the 1980s but chose to cover it up for more than 40 years.[21]

With stories like these bombarding us every day, it is easy to feel overwhelmed, complacent, and apathetic. Even if you do personally aspire to make a difference, reducing your carbon impact doesn't typically touch the heart in the same way as, say, donating money to cure cancer

---

19 Worldometers, "Current World Population". (*http://www.worldometers.info/world-population*)

20 Wikipedia, "Jim Inhofe". (*https://en.wikipedia.org/wiki/Jim_Inhofe*)

21 Suzanne Goldenberg, "Exxon Knew of Climate Change in 1981, Email Says – But It Funded Deniers for 27 More Years", *The Guardian*, July 8, 2015. (*http://www.theguardian.com/environment/2015/jul/08/exxon-climate-change-1981-climate-denier-funding*)

or volunteering at a homeless shelter might. You have to really like statistics and trust in our collective efforts to make a future difference in order for that to happen. Not everyone is hardwired to think that way.

Fossil fuel companies aren't going to stop selling oil any time soon, but we can make better choices about fuel and transportation that add up. The plummeting price of renewables has already put a noticeable dent in their market share. Companies, universities, and nongovernmental organizations (NGOs) across the planet are divesting from fossil fuels, and the tide is shifting.

Plus, individuals do find transformative ways to make a real difference. Climate Ride, a small, virtual nonprofit, is a great example. The organization empowers everyday people to collectively raise millions of dollars and bike tens of thousands of miles to help environmental organizations around the United States. Their multiday hiking and cycling events are transformative experiences for the people who participate, helping Climate Ride build a closely knit community of people banding together to make a difference. Plus, the money raised helps not one or two but more than 150 environmental organizations with key funds that drive change in sustainability, conservation, climate education, and active transportation advocacy efforts.

One Designer's Impact

To get a better understanding of an individual's impact, here's an example from someone profiled later in this book. US-based user experience (UX) designer James Christie tracked the annual carbon footprint of common tasks associated with his job in a public spreadsheet: conference travel, client visits, buying a latte, driving to work, Post-it use, book purchases, and even email and webinars attended.[22] Turns out that James produces about 20 tons of CO_2e annually just by going to work, as do most Americans (in developing countries, that number is around three tons).

---

22 See "Work and UX Activities Measured in CO2e". (*https://docs.google.com/spreadsheet/ ccc?key=0ApAANR80NbMVdEMxOXNiS1dDZkNjemlmOGdRMVg5VVE&usp=sharing*)

So should James stop going to work? Of course not. Could he explore more environmentally friendly ways to accomplish the same things in his day-to-day work life, such as telecommuting or riding his bike? Absolutely.

Knowing and understanding your individual impact is a great first step toward making changes that collectively add up. James also included a reduction column in his spreadsheet so that he can compare current behaviors with potentially more sustainable alternatives.

This also doesn't mean you should live in a yurt without power either. Your lifestyle choices are just that: choices that you should personally feel comfortable with. This book doesn't intend to shame you out of sharing cat videos; rather, it's meant to help you make more educated decisions about online conduct. It is through our collective actions that we can make a difference. Or not. It's up to us.

Current activity	CO2e cost	Source	Frequency	Annual CO2 in KG	Reduction method	New CO2e cost	New Annual CO2 in KG	Savings (Kilograms)
Lifestyle								
A latte	343g	HBB	2 a day, 250 days a year	171.5	Switch to Black coffee	23g	11	160.5
Drive to work 5 miles	11kg		2 a day, 250 days a year	5500	Cycle to work half the time		2750	2750
New iPhone	16kg			16	Get a new phone less often (every 3 years)		5.3	10.7
Using iPhone 1hr day on a 4g connection	3.4kg	HBB	365	1250	Limit 4g use to 10min a day		208	1042
An ipad every other year	130kg	Apple	0.5	75	An iPad last 3 years instead		43	32
In the office								
A paper UX book	1 kg		10	10	Read ebooks instead (no data available)		1	9
New iMac every other year	720kg		0.5	360	Replace your computer every 4 years, not 2	720kg	180	180
Postits- 2kg of these a year	2kg	http://www.twoside s.info/Pap er-Has-A-High-Carbon-Footprint		2	Use eco-postits, and fewer postits		0.3	1.7
New pens (sharpies)	84g	http://www.eco-pens.com/papermate-earth-write	25	2.2	Be thriftier with pens, and use refills		1	1.2
1 hr on computer (laptop much more energy efficent than tower PC)	69g		10 hr x 312 days	169	20% reduction in hours.		135	34
1 hr online	55g		4 hr x 312 days	68.8	30% reduction		47	21.8
1 web search	4.5g							0
1 hr webex								0
1 day in office					Work from home 1x a week saves 1.2lb (weak evidence)	544g	27	-27
Downloading data at 250MB/day.	136g	1GB = 13kWh = 544g	312	42.43	Try to avoid video	80g	25	17.43
Client visits								0
Driving for client visits (1 mile)	890g		60 miles x 2 x 25 a year	2670	More webexes - save 40% journeys	150g	1602	1068
Cycle 1 mile	65g							0
1 flight of 5000 miles								0
1 year email	135kg	HBB		135	Try to send fewer emails, I guess		120	15
1 hr big tv (what up)	240g							0
Conferences								
Air travel: conferences & client visits (2 a year, 3 days average)					2 conferences, nearer, by train			0
A night in a hotel	22kg		3x2	132	Airbnb instead of a hotel (no data)		80	52
Flights	2.3 tons		4 total	9200	Train for 500 miles each way	120 x 4	480	8720

FIGURE P-4.
Designing for impact: James Christie's spreadsheet outlining the environmental impact of his work

CO_2 Versus CO_2e

Throughout this book, the term "CO_2e" is used. Carbon dioxide, or CO_2, is a colorless and odorless greenhouse gas that is naturally occurring and emitted when fossil fuels, such as oil, coal, and natural gas are burned. It is not the only greenhouse gas emitted during human activity, however. Methane, ozone, nitrous oxide, and other naturally occurring gases can also play a role in disrupting our climate.

To account for this, scientists use the CO_2 *equivalent*—or CO_2e—to measure all greenhouse gases based on their GWP, or global warming potential. Methane, for example, has a GWP of about 25, which means that every ton of methane emitted is equivalent to 25 tons of CO_2.

For example, the Porter Ranch methane leak that occurred in early 2016 just outside of Los Angeles, California, emitted an estimated 100,000 tons of methane into the air over the course of 112 days, according to researchers. The CO_2e measurement for that leak would be 2,500,000 tons, the worst in US history.[1]

---

[1] Emma Foehringer Merchant, "The Porter Ranch Methane Leak Was the Worst in US History", *Wired*, February 25, 2016. (Source: http://www.wired.com/2016/02/porter-ranch-methane-leak-worst-us-history)

CLICK CLEAN

Greenpeace USA produces an annual report called *Clicking Clean*—focused squarely on the Internet and information and communications technologies (ICT)—that details examples of the impact our personal choices have on the environment and world energy consumption.

One of the stats in the 2014 report noted:

> If the Internet were a country, it would be 6th largest user of electricity behind China, the US, Japan, India, and Russia.

That's a lot of electricity. And the data is from 2011. Think about how much the Internet has grown since then. When we consider this, one can see that our tweets, Facebook updates, blog posts, websites, and so on, are pumping large amounts of CO_2e into the atmosphere.

FIGURE P-5.
Clicking Clean, Greenpeace's annual report on the environmental impact of the Internet, is a treasure trove of helpful data and mind-boggling stats that puts this issue in perspective

Then there's *Tweetfarts*. Based on their research, a single tweet—about 200 bytes worth of data—generates about as much CO_2e as a human fart. As of this writing, Twitter uses only 10% renewable energy to power its servers, according to Greenpeace's Click Clean Scorecard app. With 500 million daily tweets in 2012, Twitter alone generated 10 metric tons of CO_2e every day. Tweetfarts uses humor to put this impact into perspective. If you look at the Internet at large, including video, voice, and other cloud-based services, that number jumps to 830 millions tons of CO_2e annually.[23]

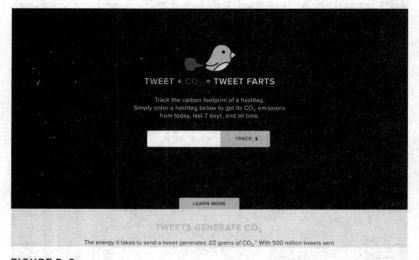

FIGURE P-6.
Tweetfarts: one tweet generates about as much CO_2e as a human fart

Although the majority of emissions come from data centers, a portion of the CO_2e released by the Internet happens on the frontend, too. A Harvard study estimated that content-heavy news sites can release more greenhouse gases than their printed counterparts if their pages are left open for extended periods of time.[24] And who hasn't left their browser tabs open in the background while they work?

---

23 American Chemical Society, "Toward Reducing the Greenhouse Gas Emissions of the Internet and Telecommunications", ScienceDaily, January 2, 2013. (http://www.sciencedaily.com/releases/2013/01/130102140452.htm)

24 Pete Markiewicz, "Save the Planet Through Sustainable Web Design", Creative Bloq, August 17, 2012. (http://www.creativebloq.com/inspiration/save-planet-through-sustainable-web-design-8126147)

Streaming video services like Netflix or Hulu play a big role, too. Imagine how many people stream the new season of *House of Cards* or *Orange Is the New Black* when they come out. In the first quarter of 2015 alone, Netflix users watched a collective 10 billion hours of streaming media.[25] Netflix and YouTube accounted for over half of peak Internet traffic in North America as of 2013.[26]

Given how quickly the Internet is growing, we're looking at a *billion tons of annual CO_2e emissions* in the very near future.

BUT ISN'T VIRTUALIZATION A GOOD THING?

For many, the most promising answers to solving the climate crisis lie in technology, be that clean tech or through the process of converting resource-intensive physical products to more lightweight online services. American entrepreneur, investor, and software engineer Marc Andreessen famously said in a 2011 *Wall Street Journal* article that "software is eating the world" and noted that "we are in the middle of a dramatic and broad technological and economic shift in which software companies are poised to take over large swathes of the economy."[27]

The ability to deliver software on a global scale has led to a digital/information economy that the United Nations estimates is in excess of $15 trillion for business-to-business (B2B) and another $1.2 trillion for business-to-consumer (B2C) with both expanding rapidly (the latter faster than the former). And that is just ecommerce. The UN's 2015 *Information Economy Report* doesn't take into account the amount of money *saved* by companies as they convert resource-heavy physical products to online services through the process of *transmaterialization* (a term we'll cover in Chapter 5).[28]

---

25 Julia Greenberg, "Netflix Says Streaming Is Greener Than Reading (or Breathing)", Wired, May 28, 2015. (http://www.wired.com/2015/05/netflix-says-streaming-greener-reading-breathing)

26 Joan E. Solsman, "Netflix, YouTube Gobble Up Half of Internet Traffic", CNET, November 11, 2013. (http://www.cnet.com/news/netflix-youtube-gobble-up-half-of-internet-traffic)

27 Marc Andreessen, "Why Software Is Eating the World", *Wall Street Journal*, August 20, 2011. (http://www.wsj.com/articles/SB10001424053111903480904576512250915629460)

28 United Nations Conference on Trade and Development, "Information Economy Report 2015". (http://unctad.org/en/PublicationsLibrary/ier2015overview_en.pdf)

In the Q4 2015 edition of *Digitalist* magazine, authors Kai Goerlich, Michael Goldberg, and Will Ritzrau researched six industries—utilities, transportation and logistics, manufacturing, retail and consumer products, agriculture and food production, and construction.[29] Based on a study by the Global e-Sustainability Initiative (GeSI) and Accenture, the authors estimated that these six industries could collectively cut 7.6 *gigatons* of emissions by using technology to digitize business processes and data to drive decisions about resource use.[30] That's a huge reduction! But it doesn't necessarily solve the problem. All those digitized business processes require hosting in data centers. They all are accessed by users with tablets, laptops, and smartphones. *All* these things run on electricity.

Meanwhile, as politicians grapple with posturing and horse trading over *who* is responsible for *what* when it comes to climate change, companies large and small are working on a wide range of options like carbon capture, biofuels, power storage, hydrogen fuel cells, cheaper renewables, and so on. At the same time, scientists focus on more speculative technologies, like cold fusion, all in the name of maintaining steady economic growth while also cutting carbon emissions. As journalist Eduardo Porter notes in *The New York Times*, "We might be able to pull it off. But it will take an overhaul of the way we use energy, and a huge investment in the development and deployment of new energy technologies."[31]

It is inspiring when large tech companies such as Google and Facebook choose to power their data centers by renewables, but what about the rest of us? Tech companies have taken the lead on this front, but they are not the only ones responsible for the Internet's impact on energy. For example, according to research by online marketing platform Agency Spotter, there are about 560,000 agencies worldwide, 120,000 of which

---

29 Kai Goerlich, Michael Goldberg, and Will Ritzrau, "Is Digital Business the Answer to the Climate Crisis?", Digitalist, November 18, 2015. (*http://www.digitalistmag.com/resource-optimization/2015/11/18/is-digital-business-the-answer-to-the-climate-crisis-2-03765081*)

30 Global e-Sustainability Initiative, "GeSI SMARTer2030". (*http://smarter2030.gesi.org*)

31 Eduardo Porter, "Blueprints for Taming the Climate Crisis", *New York Times*, July 8, 2014. (*http://www.nytimes.com/2014/07/09/business/blueprints-for-taming-the-climate-crisis.html*)

are in the United States.[32] This includes design firms, advertising companies, marketing and PR agencies, software developers, and so on. People at these agencies make decisions every day about how to design, build, and host hundreds of millions of web pages. In conducting interviews for this book, I found that very few of the agencies or freelancers I spoke to consider sustainability or renewable energy when making decisions about their work.

Often, instead of focusing on efficiency and clean power, agencies create slow, unreliable digital products and choose cheap hosts like GoDaddy (13 million customers) or HostGator (9 million customers) to house their pages. Similar to other industries, like fast food or fast fashion, buying products and services from these providers is an unhealthy decision that supports a broken system. The lack of transparency in the web hosting–only industry also makes it very easy to *greenwash*. If all these agencies had easy, affordable access to reliable green hosting services, there is an opportunity to mitigate far more emissions than the 7.6 gigatons mentioned earlier.

Greenwashing Defined

When a company spends more time touting its environmental benefits than it does implementing practices that actually make a difference, it is called *greenwashing*. "It's whitewashing, but with a green brush," say the folks at Greenwashing Index.[1]

It happens in advertising all the time. Companies make claims of a particular product or service being green to entice customers, but upon further inspection the company overall has a terrible environmental record. Greenwashing.

Specific examples of greenwashing occur throughout this book.

---

1 Greenwashing Index, "About Greenwashing". (*http://greenwashingindex.com/about-greenwashing*)

---

32 Quora, "How Many Web Agencies Are There in the World in 2014?". (*https://www.quora.com/How-many-web-agencies-are-there-in-the-world-in-2014*)

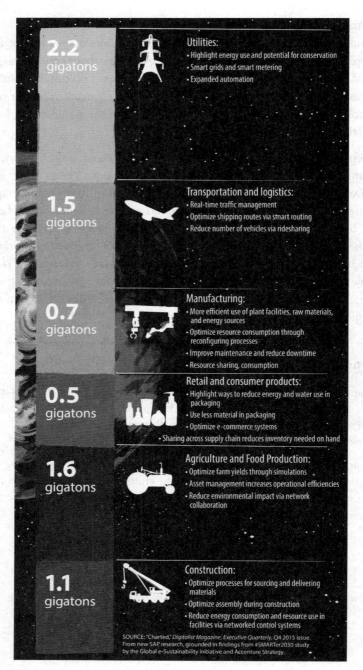

FIGURE P-7.
Digitalist magazine claims that we can cut huge amounts of emissions by moving to digital—but all those digitized services still require electricity to run

Every single one of these technology-based solutions—whether business facing or consumer facing, whether created by agencies, startups, or scientists—will have users who need to access information quickly and efficiently. They will run on servers that require redundancy of data and electricity 24/7/365. These tools should be built with the same attention to our planet's future as the solutions they intend to provide. Pixels require electricity. Thankfully, we have seen unprecedented adoption of renewables by tech companies, but we still have a very long way to go. Creating a more sustainable future for the digital economy is the problem this book addresses.

Renewables versus efficiency

President Obama's Climate Action Plan is well known for its ambitious endeavors to avoid six billion metric tons of carbon through the year 2030, the most high-profile action of which is to enforce emissions caps on power plants, which, if all goes according to plan, will cut power-sector emissions by 32%.[33]

A far lesser-known but arguably more ambitious component of the President's plan is an energy efficiency effort known as "appliance and equipment standards." It is on track to slash three billion tons of emissions by 2030 through new standards in energy efficiency for American products. Shortly after being sworn into office, the president spoke of a "revolution in energy efficiency" and even went so far as to declare energy efficiency "sexy." Since then, as many as 39 new standards were set by the Department of Energy (DOE) for everything from refrigerators and clothes dryers to pool heaters, air conditioners, and lightbulbs—but no digital products or services.

With efficiency being a key component of the Climate Action Plan and the DOE upping the pace of creating new standards, US power demand has flattened out after decades of growth, "averting the need for new plants while saving consumers billions of dollars," according to a Politico piece by Michael Grunwald.[34] These new standards are part of a larger set of efficiency initiatives that extend to fuel efficiency

---

33 WhiteHouse.gov, "President Obama's Climate Action Plan", June 2015. (*https://www.whitehouse.gov/sites/default/files/docs/cap_progress_report_final_w_cover.pdf*)

34 Michael Grunwald, "The Nation He Built", Politico, January/February 2016. (*http://www.politico.com/magazine/story/2016/01/obama-biggest-achievements-213487?o=3*)

for heavy trucks and automobiles, and extensive efforts to green the government, including the Pentagon, resulting in the lowest federal energy use in 40 years.

Efficiency and usability should sit alongside renewables as key strategies when building more sustainable digital products and services. Sure, at a time when many websites are the digital equivalent of Humvees, it's a high aspiration, but the stakes are pretty high too.

Consider this: on July 8, 2015, technical difficulties grounded all mainland United Airlines flights, took down the New York Stock Exchange for a day, forced China to suspend trading for 45% of its stock market, and *The Wall Street Journal* went offline, as well. Although some folks told us not to panic,[35] others very clearly stated that we should worry and that the reason we should worry was simple: software sucks.[36]

"It's hard to explain to regular people how much technology barely works, how much the infrastructure of our lives is held together by the IT equivalent of baling wire," says Quinn Norton in a Medium post titled "Everything Is Broken."[37] To create a more sustainable Internet, we also need to build more forward-thinking and efficient solutions. The sections of this book on UX, content strategy, and performance optimization are driven by efficiency.

Taking all this into consideration, when it comes to our own actions and creating more sustainable digital products and services, we need to ask two important questions:

- How can we power our work with renewable energy?
- How can we make our products and services—and the user experiences they offer—as efficient as possible?

As we shall see in this book's pages, those two questions can quickly become complicated. But they also offer tremendous opportunity.

---

35 Felix Salmon, "You Don't Need to Panic About the New York Stock Exchange, or Anything Else", Fusion, July 8, 2015. (*http://fusion.net/story/163160/you-dont-need-to-panic-about-the-new-york-stock-exchange-or-anything-else*)

36 Zeynep Tufekci, "Why the Great Glitch of July 8th Should Scare You", The Medium, July 8, 2015. (*https://medium.com/message/why-the-great-glitch-of-july-8th-should-scare-you-b791002fff03#.dslmkmns2*)

37 Quinn Norton, "Everything Is Broken", The Medium, May 20, 2014. (*https://medium.com/message/everything-is-broken-81e5f33a24e1#.2362w8uqo*)

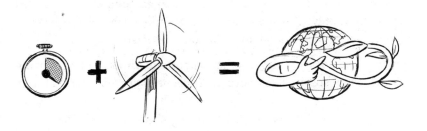

FIGURE P-8.
Efficiency and renewable energy are at the heart of everything in this book

Awareness and the consumer problem

If they consider it at all (and most don't), many people think the Internet is a green medium: after all, it replaces paper, why wouldn't it be green? However, although saving paper makes environmental sense, it might surprise many that sending an email, running a Google search, or tweeting are all actions that have an environmental impact. Plus, there is misinformation everywhere: "Discussion boards like Quora are full of 'experts' laughing at the impact of the Web on sustainability," says author and professor Pete Markiewicz. "Consumers get false information on the topic by techno-utopians."

Although the impact of a single tweet is small, when you multiply it by the number of people on this planet who send, search, and post each day—over half the world's population, by some calculations with a total of 7.6 billion expected by 2020—that impact becomes much, much larger.[38] Unfortunately, many people haven't considered the negative effect the energy used to power the Internet has on our planet.

Every photo you share, every page you browse, every movie you stream on Netflix, essentially everything you do on a computer, mobile phone, smart TV, or other device connected to the Internet requires electricity to host, serve, stream, display, and interact with. And though things are improving, the majority of that electricity still comes from

---

38 Broadband Commission for Digital Development, "The State of Broadband 2014: Broadband for All", September 2014. (*http://www.broadbandcommission.org/Documents/reports/bb-annualreport2014.pdf*)

nonrenewable resources like coal or natural gas. In fact, in the United States alone, only 13% of our power comes from renewable resources, according to the US Energy Information Administration.[39]

Raising awareness is also another important component of moving toward a more sustainable Internet. People need to know this is a thing.

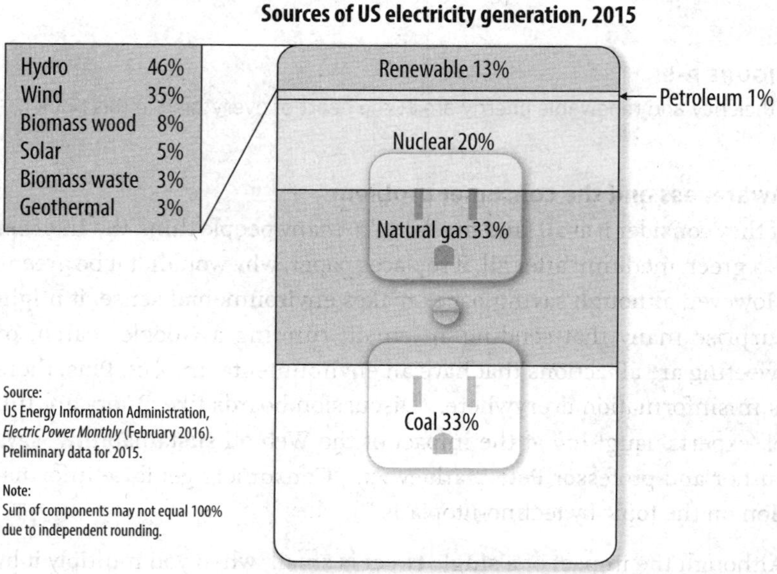

Sources of US electricity generation, 2015

Renewable 13%	
Hydro	46%
Wind	35%
Biomass wood	8%
Solar	5%
Biomass waste	3%
Geothermal	3%

Petroleum 1%
Nuclear 20%
Natural gas 33%
Coal 33%

Source:
US Energy Information Administration, *Electric Power Monthly* (February 2016). Preliminary data for 2015.

Note:
Sum of components may not equal 100% due to independent rounding.

FIGURE P-9.
In the United States, only 13% of our energy comes from renewables

B THE CHANGE: WHY I WROTE THIS BOOK

Many of the companies featured in this book are certified B Corps. These companies, which number in the thousands, are part of a global movement to redefine success in business, using the power of business to solve social and environmental problems. They range from consumer product brands such as Patagonia, Fetzer Wines, New Belgium Brewing, Ben & Jerry's, and Method cleaning products to B2B

---

39 US Energy Information Administration, "How Much US Electricity Is Generated from Renewable Energy?". (*http://www.eia.gov/energy_in_brief/article/renewable_electricity.cfm*)

companies like Rubicon Global, Arabella Advisors, Community Wealth Partners, and Cascade Engineering. Data-heavy digital companies like Etsy, Hootsuite, and Kickstarter are part of this community, as well.

In my experience, B Corps, specifically the segment of the B Corp community that focuses on digital products and services—marketing firms, web agencies, and software companies—are blazing the trail toward a more sustainable Internet. Companies like DOJO4, Open Concept Consulting, Etsy, Manoverboard, Canvas Host, Green House Data, Exygy, and others have implemented practices within their companies or otherwise pushed the idea of more sustainable digital products and services forward. Some have partnered with nonprofits like Green America or Greenpeace to promote clean cloud advocacy campaigns. DOJO4, for instance, has worked with Greenpeace to enhance its Click Clean Scorecard Chrome extension (*http://www.clickclean.org*)—which helps users better understand which of their favorite websites are powered by renewable energy.

FIGURE P-10.
Certified B Corps: people using business as a force for good

Each B Corp solves social and environmental problems in their own unique way. Certified B Corps undergo a rigorous assessment process that helps them gauge performance around accountability, transparency, social mission, and environmental impact. B Corps make profit, but they align that profit with purpose to serve *all* stakeholders, not just shareholders. This rigorous assessment—known as the B Impact

Assessment (*http://www.bimpactassessment.net*)—helps a company look carefully at each component of its business and ascertain whether all stakeholder needs are being met, such as the following:

- Are employees paid a living wage?
- Does the company recycle?
- Is profit-sharing offered?
- Does the company employ a significant percentage of minorities or people from communities with less opportunity for personal advancement?
- Is there gender equity in the company?
- Are community volunteer programs available?
- How much does the company give to charity?
- Does the company measure its energy use?
- Does the company invest in renewables?
- Does the company have a board of directors or advisory committee to help guide it through difficult business decisions?

There are hundreds of questions like the preceding ones in this assessment. Broken down into five categories—Environment, Workers, Customers, Community, and Governance—these questions guide companies down a path to building a better business that does well while also doing good.

Many questions in the Environment category focus on a company's supply chain:

- Where does the company get its supplies?
- Can waste be removed from the process of procuring, using, and disposing of supplies?
- Do vendors also make appropriate social or environmentally responsible decisions?

It was in working through the answers to many of these questions that my own thinking around how we can build a more sustainable Internet evolved.

B Mighty

My company, Mightybytes, a digital agency started in 1998 and located in Chicago, has been a certified B Corp since 2011. Becoming a B Corp is one of the best business decisions I've made and has provided our team with a clear roadmap for building a better, more conscientious company. Because many of the questions in the B Impact Assessment are related to supply chain—something I knew very little about in 2011—taking it got me thinking about what sort of environmental impact a primarily all-digital company that designs websites and creates software solutions might have. We're not manufacturing physical products. We don't source materials like fabric or plastics from offshore suppliers. At best, outside of our core employee team, the rare office supplies purchase, and occasional freelancers, we don't have a lot of suppliers. So our supply chain, I thought: green.

Except that *everything* we build requires electricity to run.

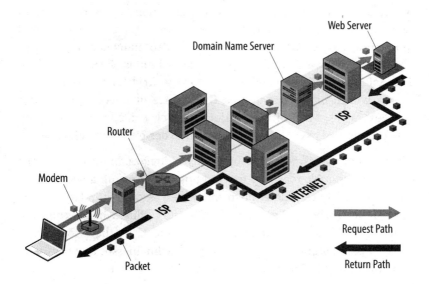

FIGURE P-11.
How green is a digital supply chain?

When the power goes out, what we build disappears. The servers that host the products and services we create require electricity 24/7/365. Some of these sites get a *lot* of traffic, which leads to electricity use in data transmission and on the frontend user side, as well. Suddenly, we realized that our little company was a bigger part of the problem than we originally thought. So, as we completed the B Impact Assessment for the first time in 2011, I began thinking about ways in which we could minimize the impact of the digital products and services we build. How could we make websites, apps, and digital marketing campaigns more energy-efficient and user-friendly?

As business, and the virtualization of business processes, is a primary force driving Internet growth, it is important for a book covering this topic to feature companies that will invest in a more sustainable future while exploring the value digital products and services can bring to their organizations. I have found B Corps to be these companies. Many of the ideas and concepts within this book were either tested on or run by fellow members of the B Corp community before implementing them as practices within our own company. They have provided valuable support and wise insights throughout the life cycle of this book.

B Corps, of course, are not the only game in town. Conscious companies and sustainable brands that pay attention to the triple bottom line—people, planet, prosperity—are cropping up all over the world to redefine success in business. There are also many nonprofits working toward more sustainable technology solutions. BSR's Center for Technology and Sustainability, The Green Grid, and specific campaigns within organizations like Greenpeace or Green America are good examples. Most of these efforts, however, focus specifically on data centers and pay little to no attention to design or frontend use.

The rigorous assessment used by B Corps provides, in my opinion, the clearest *measurable* roadmap for building and growing a socially and environmentally conscious business, which is directly related to the mission of this book. The data tells the story. Plus, when it comes to focusing on issues of sustainable design and UX, I have found that these companies are far ahead of comparable businesses that pay attention only to profit. That is why I have included so many of them in this book.

Conscious Versus Conscientious Companies

Throughout this book, the terms *conscious company* and *conscientious company* are used to describe businesses that act in the interest of people and planet alongside profit (often called *the triple bottom line*). Leaders in this movement fall into different camps about which is more appropriate.

"By definition, conscientious efforts are diligent, hard-working, careful of meeting outside expectations, ultimately defined by the world around the person or 'outside in'," says Thea Polancic, Founder and Chair of Conscious Capitalism Chicago (*http://consciouscapitalismchicago.org*). "The *conscious* in conscious capitalism both refers to a more awake or aware way to do business; but also to the development of human consciousness. A new form of leadership is emerging that reflects how we are evolving as human beings, towards more complex ways of systems thinking as well as a greater capacity for love and care."

Maren Keeley, cofounder of *Conscious Company* Magazine agrees (*http://consciouscompanymagazine.com*). "When we named our magazine, we wanted to convey the difference between many companies that act in their own interest seemingly without awareness, respect of, or concern for the environment or people, and those that are 'behaving' with awareness," she says. "To say "conscientious" to me only implies that it's about morality, which is certainly a key element, but I also believe this will make companies more resilient overall."

"While in the United States we often treat companies as though they are people," Maren says. "Technically, they are groups of people, acting together, not alone. Companies themselves are neither conscious nor conscientious," Maren says. "But my hope is that the people operating these companies can consciously be aware of the impact that their actions, and the company, have on the world and others."

Bryan Welch, CEO of B the Change Media has a different opinion: "To be conscious literally only means that we are not 'unconscious.' Venal, greedy, unprincipled companies are perfectly conscious. Conscious of their revenues and expenses. Conscious of their supply and demand. Conscious of their customers and suppliers. However, they are not conscientious." (*http://bthechange.com*).

"Worse yet," he says. "It's my observation that only people inside our 'tribe' of do-gooders recognize this meaning of 'conscious.' It is confusing to the vast majority of English speakers. And I am against speaking 'tribal' languages. So we use conscientious as a word that means that we try to do things the right way and spread the maximum amount of benefit in the world."

Who Should Read This Book

This book is primarily written for those who design, build, and manage digital products and services. Whether you are a product manager, a site owner, a UX designer, a frontend developer, a content strategist, or a unicorn who does all the above, you will hopefully find something useful within these pages.

As it covers an area often neglected in sustainability assessments, sustainability or Corporate Social Responsibility (CSR) professionals should find some useful ideas within these pages, as well. Several pros have offered their expertise throughout each chapter on topics not typically covered in standard sustainability assessments.

Here are some willful intentions for this book:

- Inspire readers to think about and be more aware of web sustainability alongside bigger sustainability issues.
- Encourage web design and development teams to connect efficiency, reliability, usability, and sustainability so that they can make better choices in their work.
- Help sustainability and CSR professionals better understand Internet sustainability so that it can be folded into their existing work.
- Provide readers with the knowledge necessary not only to be inspiring design leaders but also environmental advocates, as well.
- Finally, to inspire people who create and build the Internet to power it with renewable energy.

How This Book Is Organized

Book chapters begin and end with learning objectives. At the beginning of each chapter, expectations are set for what you will learn in that chapter. At the end, next step tasks are offered to translate concept into action.

- Chapter 1 defines sustainability and frames the conversation with statistics and research on the overall environmental impact of the Internet.
- Chapter 2 outlines a framework for designing more sustainable digital products and services.

- Chapters 3 through 6 dive deeper into each category of the aforementioned framework: design and UX, content, performance optimization, and green ingredients, including green hosting.
- Chapter 7 includes research on devising meaningful ways to measure the environmental impact of digital work.
- Chapter 8 posits what a more sustainable future Internet might be.

Sidebars, interviews, and callouts throughout each chapter highlight specific topics, definitions, or complementary content.

Let's Go!

Thank you for choosing this book. It has been a labor of love several years in the making. Now that I have told the backstory and set the stage, I hope you enjoy what follows.

[1]

Sustainability and the Internet

What You Will Learn in This Chapter
In this chapter, we will cover:

- A definition for sustainability and the role it plays in business.
- How organizations use sustainability principles to innovate, differentiate themselves, reduce waste, and function more efficiently.
- How sustainability applies to the Internet.
- How virtual life cycle assessments might help web teams devise more sustainable solutions.

A Greener Internet

"The Internet is the single biggest thing we're going to build as a species," says Greenpeace's Gary Cook in an article from *The Atlantic*.[1] "This is something that if we build it the right way, with the right sources of energy, could really help power our transition to renewables. If we build it the wrong way, it could actually exacerbate the problem."

This is a book about designing the Internet the right way: efficient, accessible, future-friendly, and powered by renewables.

BUILDING SUSTAINABLE SOLUTIONS
Throughout this book, we're going to dive into great detail about all the checkpoints we can put into place to help us make better digital products and services that not only keep users happy and engaged but that

---

1 Ingrid Burrington, "The Environmental Toll of a Netflix Binge", The Atlantic, December 16, 2015. (*http://www.theatlantic.com/technology/archive/2015/12/there-are-no-clean-clouds/420744*)

are also more efficient and save energy, as well. We will discuss how to translate these checkpoints into an easy-to-understand framework that will help you and your clients make more sustainable design and development decisions.

Rarely are the things that we set out to build the things that actually get built. In the maelstrom of changing opinions, validated learning, contractual obligations, and shifting stakeholder requests, we are expected to produce magic. To stay on time, on budget, or on good terms with the person signing our checks, we cut corners. We give in to impractical requests. We let the client autoplay their 30 MB video on the home page. We add an image carousel that over time becomes filled with a dozen photos of generic business people shaking hands. And in turn the average web page size—according to the HTTP Archive, which tracks how the Web is built—rises to more than 2.3 MB.[2]

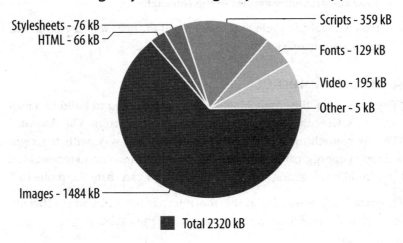

FIGURE 1-1.
The average web page size hit 2.3 MB in early 2016

---

2 HTTP Archive, "Interesting Stats". (*http://httparchive.org/interesting.php?a=All&l=Apr%201%202016*)

All that data bouncing back and forth through our networks requires electricity. Energy is used to host our content, serve our content, and interact with our content. Unfortunately, very little of that energy comes from clean or renewable resources. Hence, the Internet, which many people might think is a "green" medium due to the simple fact that it often replaces paper, isn't as clean as we might think it is. In fact it produces more greenhouse gases (GHGs) than the airline industry,[3] which produced 770 million tons of emissions in 2015.[4] With more than three billion active Internet users in early 2015 and more than half the world's population predicted to be online by the end of 2016,[5] it is well on track to produce more than a billion tons of GHGs any day now, if it isn't doing so already by the time you read this.

We don't mean to build bloated websites that clog networks, waste electricity, and frustrate users. Somehow—be it a desire to try out the latest design bell or programming whistle or the need to keep demanding stakeholders happy—it just turns out that way. If design and development teams could check themselves every step of the way by simply asking "Is this the most sustainable, efficient solution?", many of these bad decisions—which have detrimental consequences like lost revenue and frustrated customers—could be headed off at the pass. And the Internet would be a cleaner, greener, happier place for all.

But first, let's offer a bit of context.

---

3   American Chemical Society, "Toward Reducing the Greenhouse Gas Emissions of the Internet and Telecommunications", January 23, 2013. (http://www.acs.org/content/acs/en/pressroom/presspacs/2013/acs-presspac-january-23-2013/toward-reducing-the-greenhouse-gas-emissions-of-the-internet-and-telecommunications.html)

4   Air Transport Action Group (ATAG), "Facts and Figures". (http://www.atag.org/facts-and-figures.html)

5   Internet Live Stats, "Internet Users". (http://www.internetlivestats.com/internet-users)

Sustainability Defined

The word "sustainability" and its variations are thrown around often, maybe not as much as the term "green," but still quite a lot. Marketers love using it to describe their products' virtues. Environmentalists espouse a more hopeful future in its promise. In certain circles, its overuse has perhaps watered down the word's meaning and trivialized the fact that few things are truly sustainable.

The most commonly quoted definition of sustainability comes from a paper called *Our Common Future*, released in 1987 by an organization known as the Brundtland Commission,[6] named by the United Nations after the former prime minister of Norway and tasked with the mission of rallying countries to slow deterioration of our environment and natural resources:

> Sustainable development...meets the needs of the present without compromising the ability of future generations to meet their own needs.

At the time, the hope of this organization was to reconcile economic development with environmental damage. Nearly 30 years later, this dream of reconciliation between people and planet remains elusive. As Naomi Klein points out in her book *This Changes Everything: Capitalism vs. The Climate* (Simon & Schuster, 2014):

> Our economic system and our planetary system are now at war. Or, more accurately, our economy is at war with many forms of life on earth, including human life. What the climate needs to avoid collapse is a contraction in humanity's use of resources; what our economic model demands to avoid collapse is unfettered expansion. Only one of these sets of rules can be changed, and it's not the laws of nature.

Those of us who create the digital future are smack-dab in the crosshairs of this fight between economic expansion and environmental collapse. The rise of the Internet has fueled our increasing demand for products of all shapes and sizes, each of which requires substantial

---

[6] Wikipedia, "Brundtland Commission". (*https://en.wikipedia.org/wiki/Brundtland_Commission*)

resources to create, transport, use, and dispose of. As we shall see in upcoming chapters, although physical products might require the lion's share of these resources, digital products and services have their own role to play, as well.

FIGURE 1-2.
With so many doom-n-gloom environmental messages out there (some of questionable motivation) it is easy to see why people feel burnt out or overwhelmed and don't think they can make a difference

SYSTEMS THINKING AND SUSTAINABILITY

Progress has been made since the definition of sustainability was coined. Sustainability is now its own industry, driven by data and chock full of consultants, analysts, corporate social responsibility (CSR) managers, and scientists. Many sustainability professionals have adopted a systems-thinking approach to sustainability rather than view it simply as a compliance or regulation issue, which is common. These people focus on how a company's systems and its constituent parts interrelate within the larger system of the environment over time.

In describing the definition of sustainability put forth by the Brundtland Commission, the International Institute for Sustainable Development notes that:[7]

> All definitions of sustainable development require that we see the world as a system—a system that connects space; and a system that connects time. When you think of the world as a system over space, you grow to understand that air pollution from North America affects air quality in Asia, and that pesticides sprayed in Argentina could harm fish stocks off the coast of Australia. And when you think of the world as a system over time, you start to realize that the decisions our grandparents made about how to farm the land continue to affect agricultural practice today; and the economic policies we endorse today will have an impact on urban poverty when our children are adults.

Using this systems-thinking approach (and others), the sustainability industry has helped many businesses adopt more sustainable practices, resulting in reduced GHG emissions, less waste, more efficiency, and often higher profitability.[8]

Author Nathan Shedroff also covers this extensively in his book *Design Is the Problem* (Rosenfeld, 2009):[9]

> The essence of this definition [of sustainability], which may not be obvious immediately, is that needs aren't just human, they're systemic. Even if you only care about humans, in order to care for humans, you need to take care of the system—(the environment) that you live in. And this environment doesn't include just the closed system we call the planet Earth. It also includes the human systems we live in—our societies—and the forming, changing, and constantly evolving values, ethics, religion, and culture that encompass these societies. We aren't separable from each other, and we can't ignore the effects of the whole—nor should we.

---

7 International Institute for Sustainable Development (IISD), "Sustainable Development". (*https://www.iisd.org/sd*)

8 Sunmin Kim, "Can Systems Thinking Actually Solve Sustainability Challenges? Part 1, The Diagnosis", Erb Perspective Blog, Jun 4, 2012. (*http://erb.umich.edu/erbperspective/2012/06/04/systems-thinking-part-1*)

9 Nathan Shedroff, *Design Is the Problem* (Brooklyn, NY: Rosenfeld Media, 2011).

To date, few people have applied this same type of thinking to digital products and services. With all of its figures, metrics, and systems-based thinking, the field of sustainability should be a natural fit for those of us who create the nervous system of the 21st century known as the Internet. But what does it really mean to be sustainable? The reality is it's next to impossible. Few things we humans create are truly sustainable. In a 2004 *Grist* interview, Patagonia founder Yvon Chouinard noted:[10]

> There's no such thing as sustainability. There are just levels of it. It's a process, not a real goal. All you can do is work toward it. There's no such thing as any sustainable economy. The only thing I know that's even close to sustainable economic activity would be organic farming on a very small scale or hunting and gathering on a very small scale. And manufacturing, you end up with way more waste than you end up with finished product. It's totally unsustainable. It's just the way it is.

In the green building movement, some have suggested that we move beyond thinking about sustaining an already degraded planet and more toward regenerative design rather than simply sustainable design. In other words, defining success simply by being impact-neutral is not enough. As a species, we need to renew and regenerate, to change things for the better. According to the folks at Thrive Design:[11]

> This entails going even deeper than the leading edge of systems thinking (the uncovering of the interconnected and complex nature of the world) and into the realm of systems *being*. It recognizes that we *are* the system, that there is and can never be any separation of humans from the web of life that we wholly depend upon. When we release toxins into the environment, we quickly discover them infiltrating our own bodies through the food we eat, the water we drink, and the air we breathe. In essence, what we do to nature, we do to ourselves. This perspective can help us come to see the planet and its life supporting systems as part of our 'extended body'. A natural response to

---

10 Amanda Little, "An Interview with Patagonia Founder Yvon Chouinard", Grist, October 23, 2004. (*http://grist.org/article/little-chouinard*)

11 Joshua Foss, "What is Regenerative Development?", Thrive Design Studio. (*http://www.urbanthriving.com/news/what-is-regenerative-development*)

this understanding is of care and compassion as it becomes our own self-interest to actively manage the health and integrity of the living systems that we rely upon.

This switch from focusing on outcomes to focusing on process is one that fits well within the context of this book. Startups, digital agencies, and software companies are consistently moving toward models of continuous deployment, where the features and functions of product releases are ongoing and the systems built are actively managed much like those mentioned before. When we talk about designing greener digital products and services, we do so with the idea that the Internet will never be a truly sustainable place. It will always use resources. There will always be work to do. If you have ever designed a website or mobile app, you know that meeting the needs of the present without compromising those of the future is a monumental challenge indeed. After all, when was the last time you played a Flash game on your phone?

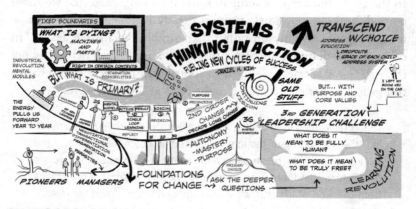

FIGURE 1-3.
Systems thinking requires that we see the world as a system that connects space and time

Sustainability in Business

Designing greener digital products and services requires us to better understand the role of sustainability in business as a whole, for it is into these systems and processes where we will plug our own work. Although it might not be the norm today, assessing the impact of a company's digital products and services will hopefully be just one common future component of greater sustainability initiatives in business.

The ways in which sustainability concepts are used in business are so broad they can be hard to encapsulate in a single chapter overview. Different companies take different approaches, use different tools, and focus on different things based on the resources available to them and what they want to achieve from their efforts. Each organization tends to chart its own course using its resources, tools, and goals as drivers. Some focus on energy efficiency, others on waste reduction. Some companies are driven by marketing goals, whereas others use principles of sustainability to drive innovation and disrupt industries. Some focus on a triple bottom line that pays equal attention to people, planet, *and* profit. The best-in-class do all of the above.

Cradle to cradle (C2C), for instance, is a biometric approach to the aspiration of waste-free products and systems design. It models business and manufacturing practices on natural processes and suggests that industry must protect and enrich ecosystems and nature's biological metabolism while also maintaining a safe, productive, technical metabolism for the high-quality use and circulation of organic and technical nutrients.[12] In other words, rather than producing waste, the traditional waste products at the end of a product's life cycle are reused to give life to something new or are reintroduced to the life cycle of the original product, creating what is commonly known as a "closed loop" system, where outputs are reintroduced as inputs. Fairphone, a Dutch company that sources e-waste from Ghana for use in its smartphone devices, is a great example of C2C in practice. More on Fairphone in later chapters.

---

12 Michael Braungart and William McDonough, *Cradle to Cradle: Remaking the Way We Make Things* (New York: North Point Press, 2002).

FIGURE 1-4.
Data drives sustainability principles at many businesses

Cradle to grave, on the other hand, refers to a company taking responsibility for the disposal of goods it has produced and is most commonly associated with life cycle assessments, which we cover in the following section.[13] C2C approaches are generally more favorable than cradle to grave due to the closed loop system concept mentioned above.

Some companies use a helix of sustainability, another systems-based approach which also maps models of raw material use onto those of nature. The helix guides six categories of key business functions (governance, operations, corporate culture, process, marketing, and

---

13 The Dictionary of Sustainable Management, "Cradle-to-Cradle". (*http://www.sustainabilitydictionary.com/cradle-to-cradle*)

stakeholders) through five levels of commitment to sustainable principles (no commitment, exploration, experimentation, leadership, and restoration).[14]

FIGURE 1-5.
The helix of sustainability

Factors such as number of employees, size of supply chain, financial resources, company mission, and product and service types also play important roles. Some have internal departments to manage these affairs while others hire consultants or firms to assess impact and make recommendations for them.

Because reducing an organization's existing environmental impact and devising processes for minimizing future environmental impact can take so many forms, let's explore some of the ways in which this can play out:

- Identifying (and, later, implementing) efficiencies
- Life cycle assessments (LCAs)
- Benchmarking
- Marketing, branding, and greenwashing
- Innovation and disruption

Although some of these processes do include assessments of electricity used by internal data centers and employee workstations, they rarely take into account energy used by websites, mobile apps, cloud-based services, or other digital products and services. This offers unique challenges and opportunities.

---

14 The Dictionary of Sustainable Management, "Sustainability Helix". (*http://www.sustainabilitydictionary.com/sustainability-helix*)

IDENTIFYING EFFICIENCIES

Many companies start down the road to greater sustainability by identifying efficiencies that let the organization reduce waste of any form while also saving money. This is the low-hanging fruit to sustainability professionals. These could be efficiencies in electricity use, productivity, process improvements, and so on. This is a practice particularly common in larger organizations, as saving money is an easy sell to shareholders and decision makers. Unfortunately, when organizations just care about cost, they might only identify the lowest-hanging fruit and not do more because the payback is longer for higher-hanging fruit (instead of bundling a mixture of both).

Identifying efficiencies is often part of the larger goal and scope process of a life cycle assessment, discussed in more detail a bit later. To accurately identify efficiencies, companies must first accurately identify sources of waste. Companies need good (and ongoing) measurement for this, which might be easier for energy but more expensive and/or challenging for physical waste.

By assessing the entire life cycle of their products and services, from cradle to cradle, as noted earlier, companies can identify sources of waste, and in turn, identify corresponding efficiencies to improve performance. This is a common workflow in sustainability practices, yet to date very few companies that offer sustainability services or life cycle assessments apply this process to digital products and services. A framework for doing so will be presented later in this chapter and discussed throughout the book.

LIFE CYCLE ASSESSMENTS

Life cycle assessments (LCAs) are commonly used to calculate the environmental impact of a product or service during its entire lifetime. Also sometimes referred to as *cradle-to-grave*, or more recently, *cradle-to-cradle* assessments, as noted earlier, they entail multiple steps:

1. Goal and scope definition (What are we trying to accomplish?)
2. Inventory analysis (What are we trying to assess?)
3. Impact assessment (What is the impact of our inventory?)
4. Interpretation (What does the data tell us?)

After these steps have been defined, a rigorous process begins to help the company or organization better understand its environmental impact and then devise a plan to do something about it.

To apply these concepts to digital products and services you might answer the preceding questions as follows:

Goal and scope definition
    Define the environmental impact of our online properties and implement a plan for mitigating that impact.

Inventory analysis
    Web, apps, cloud services, social media, and so on—what is included and what isn't?

Impact assessment
    How much CO_2e is generated by these properties? Is there other waste, as well?

Interpretation
    Create a plan to reduce impact—renewable energy, carbon offsets, increases in product/service efficiency, e-recycling, and so on.

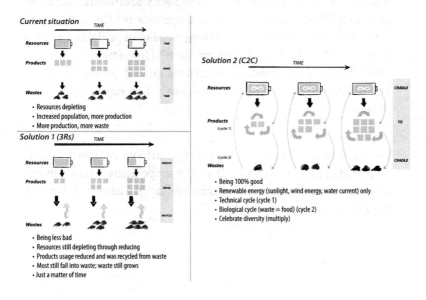

FIGURE 1-6.
Life cycle assessments take a product or service's impacts into account from cradle to grave, or better yet, from cradle to cradle

Setting goals and scope

It is important with any complex project to set specific goals, timeline, budget, scope of work, and so on. However, enough flexibility should be built into a project's scope to allow for iteration, experimentation, and the ability to pivot from a specific process should it not yield fruitful results.

We will talk about Agile workflows in later chapters, because they relate to creating more sustainable digital products and services, but it is important to note that building iteration and collaboration into the process of any project is often a more sustainable approach that bears better results than those that require extensive scoping up front with little room for flexibility after a project is in process.

If, for example, the focus of your efforts is solely on greenhouse gases, your scope might look like this:[15]

Set boundaries

> Define an inventory's physical, virtual, organizational, and operational boundaries. Traditional LCAs are rarely full scope (cradle to cradle) and the assessment can only begin when materials arrive at a company's gate. Power use for designer workstations, for example, would fall under the scope of a traditional LCA, but if applying this to digital, are you including website or social media, too? If it is in the cloud, is it your problem? How much do you care about frontend, user-driven electricity during product use? For additional considerations, see the sidebar "The Power of Users" on page 15.

Define scope

> Decide which emissions source and/or activity categories should be included in the inventory. Does the scope include both physical and virtual properties? The details you define here will drive the level of effort required.

---

[15] US Environmental Protection Agency, "Climate and Energy Resources for State, Local, and Tribal Governments". (*http://www3.epa.gov/statelocalclimate/state/activities/ghg-inventory.html*)

Choose quantification approach
  Depending on the data available and the purposes of the inventory, define how you will approach collecting data and where it will come from. In a traditional LCA, this might be identifying a source for emissions factors (EFs) and a formula for measuring them.

Set a baseline
  When choosing a baseline—the point that marks the beginning of your measurement efforts—to benchmark progress going forward, consider the following:

  - Whether the data is available for the time you've chosen
  - Whether the chosen timeframe is representative
  - Whether the baseline is coordinated to the extent possible with baseline years used in other inventories (if they exist)

Engage stakeholders
  Bring stakeholders into the inventory development process early on to provide valuable input on establishing a baseline. If a specific department or position will be affected by the data derived from this process, it will help to have them on board from the beginning.

Consider certification
  Consider a third-party certification such as ISO 17024, EPt(GHG), or those offered by CSA Group. This will ensure quality and that your inventory is complete, consistent, and transparent.

The Power of Users

Although you can't control what end users put on their phone, how they treat their battery, or whether they keep their laptops running at peak efficiency, you *can* make sure you're not part of the problem. As we'll see throughout this book, slow-loading digital products and services that are unreliable have a real impact on the bottom line as well as on your reputation with customers. But is their power use something you should measure? Isn't that *their* problem?

If you can measure how much use your application gets across devices and platforms, you can estimate how much electricity it needs. This is a standard LCA concept that can be applied to digital. If you can measure it, you can reduce it by using many of the optimization techniques we'll discuss in the chapters ahead. You can also offset it, which we'll discuss momentarily.

Inventory Analysis

This is where data collection happens. Life cycle inventories (LCIs) can be complex affairs. Analysts track all of the inputs and outputs to a business system, including (but not limited to) raw materials; energy use; emissions to air, water, and soil (tracked by substance); and so on. The complexity of a company's supply chain is one of many factors that will dictate the complexity of the inventory.

In Internet parlance, this would be what is commonly referred to as analytics. A popular tool like Google Analytics, for example, offers you all the things you *can* measure—with the notable exception of electricity use. In the name of efficiency, it is up to you to take an inventory of the metrics you *should* measure.

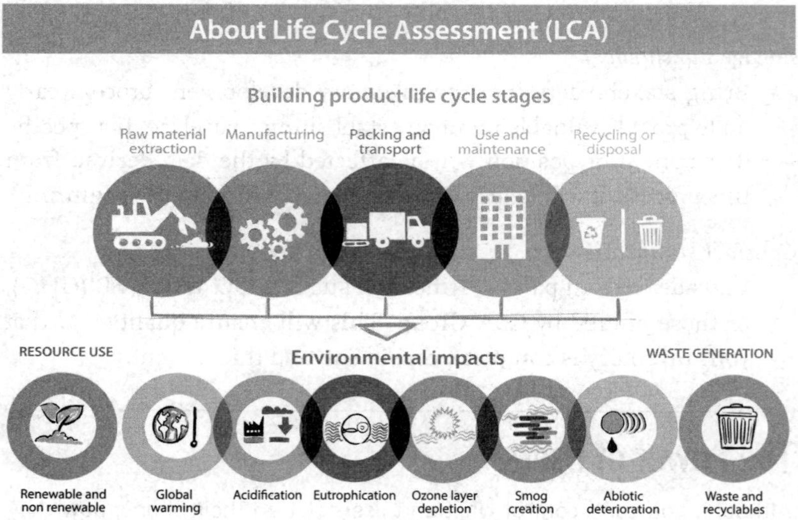

About Life Cycle Assessment (LCA)

Building product life cycle stages

Raw material extraction · Manufacturing · Packing and transport · Use and maintenance · Recycling or disposal

RESOURCE USE — **Environmental impacts** — **WASTE GENERATION**

Renewable and non renewable · Global warming · Acidification · Eutrophication · Ozone layer depletion · Smog creation · Abiotic deterioration · Waste and recyclables

FIGURE 1-7.
Life cycle assessments can help you measure what really matters in your organization

For reference, common components of *standard* (i.e., physical products) LCIs include materials, manufacturing, packaging, distribution, usage, and disposal. We'll discuss how to apply these components to digital products and services later in this chapter. For each of these components, consider the following questions:

Materials
> What are the materials used to make the product? Where are they sourced from? How much energy is used in that process? How much emissions are produced in both sourcing and creating the materials? How can they be made or sourced in a more sustainable manner?

Manufacturing
> How much energy does the manufacturing process use? How much waste does it produce? Where can you reduce emissions in the process?

Packaging
> What materials are used in the product's packaging? Does the product use soy-based inks or compostable materials, for example, versus a less sustainable equivalent? How much energy is used in the packaging process? What about waste? Can that be reduced at all?

Distribution
> How much energy is used in the process of distributing the product? Does it need to be shipped in a cargo container from China or can it be walked down the street to the store where it is sold? How much waste or emissions are produced when distributing the product?

Usage
> Does the product require energy when being used? Is waste produced? How can product usage be more efficient in energy consumption? How can it produce less waste?

Disposal
> When the product is disposed, can its materials be reused? Are there ways in which it can produce no waste at the end of its life cycle?

Impact assessment

This process, also part of a full LCA, is where the data collected is analyzed and the impact of a company's product or service assessed. In other words, a certain manufacturing process might require a specific amount of oil or natural gas, something that is typically included as part of the inventory that we just discussed. The impact assessment will decipher the environmental effect of that process. For digital products and services, an assessment will use data pulled from analytics tools to gauge a product or service's environmental impact.

Interpretation

Finally, all that data needs to be interpreted and a plan of action put into place to decrease environmental impact. This includes sourcing renewable power, increasing energy efficiency, reducing waste, and in general making the company's supply chain—including its digital products and services—more sustainable.

The Greenhouse Gas Protocol

An important part of an LCA is calculating GHG emissions. As emissions from electricity use can comprise the majority of a digital product or service's waste, this is critical to successfully analyzing their impact. The Greenhouse Gas Protocol from the World Resources Institute serves as the foundation for nearly every GHG standard and program in the world. It is the most widely used international accounting tool for government and business leaders to understand, quantify, and manage greenhouse gas emissions across a product's life cycle.

Product life cycle emissions are all the emissions associated with the production and use of a specific product, from cradle to grave, including emissions from raw materials, manufacture, transport, storage, sale, use, and disposal.

The Greenhouse Gas Protocol can help organizations do the following:[16]

- Determine and understand risks and opportunities associated with value chain emissions.
- Identify GHG reduction opportunities, set reduction targets, and track performance.

---

16 Greenhouse Gas Protocol, "FAQ". (*http://www.ghgprotocol.org/files/ghgp/public/FAQ.pdf*)

- Engage suppliers and other value chain partners in GHG management and sustainability.
- Enhance stakeholder information and corporate reputation through public reporting.

It is also important to note that the GHG Protocol only focuses on greenhouse gases. It is not meant to show an indication of a product's entire environmental impact, which could include other kinds of waste. Because electricity powers everything on the Internet, its biggest source of waste is GHG emissions, but as we will cover in later chapters, the hardware used to create, serve, and interact with web content produces other forms of waste, too.

Although the Greenhouse Gas Protocol is currently the gold standard, it is also worth noting that the Sustainability Accounting Standards Board (SASB) has a mission to devise sustainability standards for different industries based on accounting standards but for environmental, social, and governance (ESG) issues as opposed to financial performance.

BENCHMARKING

It should be noted, too, that an important component of sustainability in business is benchmarking improvement. How do you know if you're improving if you aren't continuously measuring results and comparing them against the last time measurements were taken as well as against competitors? It is important that companies dedicate resources to these endeavors in amounts appropriate for the organization. Otherwise, initial efforts are wasted and improvement is minimal.

Given the similarity in process between this and many digital endeavors—such as Agile and iterative design strategies, website performance optimization, or digital marketing campaign measurement—one might think there would be more synergy between technology, design, marketing and sustainability departments, but with few exceptions this isn't often the case.

Which brings us to our next section.

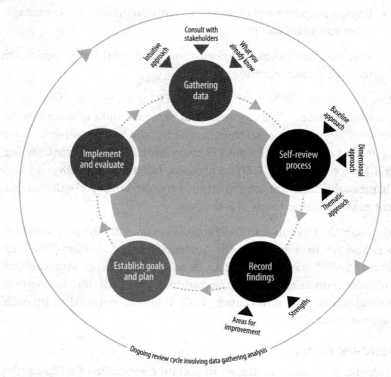

FIGURE 1-8.
Benchmark it: an important concept in this book.

MARKETING, BRANDING, AND GREENWASHING

Some companies pursue sustainability initiatives to meet marketing objectives or to appease shareholders, customers, suppliers, and so on. If the company doesn't take its environmental initiatives seriously, this can offer opportunities for what is commonly referred to as *greenwashing* (discussed in "Greenwashing Defined" on page xxiii). Fossil fuel companies might, for example, spend large sums of money on campaigns about how they're "going green" even though the product they sell is the biggest contributor of GHG emissions. Examples include "clean coal," "clean natural gas," and Volkswagen's "Dieselgate" emissions scandal (VW engineers intentionally programmed diesel engines to activate certain emissions controls only when the vehicle was undergoing emissions testing, while in real-world driving, the vehicles' nitrogen oxide output was up to 40 times higher than during tests and were less efficient on performance and fuel efficiency than what the company claimed).[17] Coca-Cola, Air France, BP, and many others have come under fire from environmentalists for touting their commitment to the planet while also polluting or putting people and communities at risk.[18]

"In reality, every company with a green claim is greenwashing," says sustainability consultant JD Capuano. "Our current economic system is based on unlimited growth, but our resources are limited. Even a company that makes the most responsible products can still have significant environmental impact."

This is often a design problem as well. Because marketing campaigns are so heavily reliant on good design to communicate their message, designers are often hired to greenwash, and in some cases they might even consider themselves "green" because of their work when in reality that is not the case.

It is up to the individual designer to define where he draws the line in the sand, and for many—especially those who freelance for a living, where a single project could make or break your monthly finances—that can be a moving target. In the early days of my career, I was hired for a motion graphics project that marketed a certain type of cigarette as an active lifestyle brand. Yup, you read that right: snowmobiles, skiing, extreme winter sports, and cigarettes apparently all go hand-in-hand. At the time I really needed the work so I took the gig, but in retrospect it's one of several freelance projects that, given the chance, I would reconsider. Projects like these inspired me to focus more heavily on doing mission-driven work that mattered as Mightybytes grew and evolved.

Given the lack of consistent, common standards for what CSR or sustainability initiatives should entail and how truly good companies should perform, it can be difficult to discern between a good company and just good marketing. Some companies—sustainable brands, B Corps, and those that subscribe to the philosophy of conscious capitalism, for instance—are able to align profit with purpose in ways that

---

17 Robert Duffer and Tribune staff, " Volkswagen Diesel Scandal: What You Need to Know" Chicago Tribune, September 22, 2015. (http://www.chicagotribune.com/classified/automotive/ct-volkswagen-diesel-scandal-faq-20150921-story.html)

18 Breena Kerr, "The Culprit Companies Greenwashing Climate Change", The Hustle, December 6, 2015. (http://thehustle.co/the-culprit-companies-greenwashing-climate-change)

help the company thrive financially while also solving social or environmental problems. By undergoing the rigorous B Impact Assessment, however, only certified B Corps prove that they adhere to higher standards of accountability and transparency, giving consumers confidence that their products and services haven't been greenwashed. We will discuss this in more detail in Chapter 3 when we cover triple-bottom-line business models.

FIGURE 1-9.
Greenwashing: it might sound nice, but don't look under the hood

INNOVATION AND DISRUPTION

Some leaders in this movement are disrupting the status quo with approaches like biomimicry, a process that seeks sustainable solutions to human challenges by emulating the patterns and strategies found in nature. New York–based Ecovative Design,[19] for instance, grows packing and building elements from biodegradable mushroom-based materials (see Figure 1-10) rather than manufacturing them from more environmentally toxic sources. These products can replace polystyrene, a more hazardous material.

---

19 Ecovative Design, "How It Works". (*http://www.ecovativedesign.com/how-it-works*)

FIGURE 1-10.
Ecovative Design creates packaging and building materials from mushrooms.

Or take Nascent Objects,[20] a more sustainable electronics platform that could disrupt how consumer electronics are designed, manufactured, and sold (see Figure 1-11). Their system of modular electronics components operates on the premise that filling landfills with outdated gadgets or purchasing the latest edition of a device just for a single feature is wasteful. The company's business model is shaking up the way consumer electronics have been brought to market for the last 60 to 80 years. Rather than releasing single-use products that lock the electronics inside, they have created a system of 15 or so common electronics modules—speaker, camera, microphone, and so on—and made them interchangeable. The result is a system with untold potential and a company whose mission is to create a sustainable electronics revolution.

Choosing from various modules, customers can use the company's software to create electronics devices specific to their needs, such as baby monitors, WiFi speakers, water meters, security cameras, and more. The company uses 3D printing to create the device's housing and ships the product to the customer. If a product is outgrown—such

---

20 Patrick Sisson, "Nascent Objects, a Sustainable Electronics Platform, Wants to Make Gadgets More Green", Curbed, March 11, 2016. (http://www.curbed.com/2016/3/11/11201000/nascent-objects-modular-electronics-sustainable)

as a baby monitor—customers can reuse the internal modules to create something new. So far, the company has designed a water meter, a wireless speaker, and a security camera with many more products in the pipeline

FIGURE 1-11.
Nascent Objects is disrupting the way consumer electronics are designed, manufactured, and sold

Company founder Baback Elmieh believes that an increase in the number of distributed supply chains coupled with better technology lowers the barrier to entry for designers. Using a system like the one offered by Nascent Objects means that startups can now release products that previously required the work of a much larger company.

It is unclear what happens to device housings once the product is no longer of use—are they sent back to Nascent Objects to be reused? Plus, 3D printing has its own environmental implications: 3D printers can waste about 40% of nonrecyclable materials and because they require keeping plastic materials melted in order to work, they use a lot of electricity.[21] A design shop that keeps them running all day could ostensibly run up a large carbon footprint for each printed piece. Plus, a study from early 2016 showed that some 3D printers emit styrene particles, a possible carcinogen.[22]

---

21 Adele Peters, "Is 3-D Printing Better for the Environment?", Co.Exist, January 29, 2014. (http://www.fastcoexist.com/3024867/world-changing-ideas/is-3d-printing-better-for-the-environment)

22 Steve Dent, "Study Shows Some 3D Printing Fumes Can Be Harmful", Engadget, February 1, 2016. (http://www.engadget.com/2016/02/01/study-shows-some-3d-printing-fumes-can-be-harmful)

Still, given the huge environmental impact of large-scale electronics manufacturing, Nascent Objects' model offers enormous potential to minimize impact by rethinking the role of devices in our lives, offering solutions with reuse built right into the business model.

Similarly, sensors associated with Internet of Things (IoT) applications offer vast opportunities to monitor energy use and control energy-producing hardware like solar panels. A Nest thermostat, for example, provides consumers with accurate real-time data on energy use, helpful monthly reports, and the ability to turn down or shut off your HVAC system when you're not home, resulting in significant savings in both energy and money. There are many business applications for IoT tech, as well. We have only just scratched the surface here.

LCAS IN ACTION

In 2014, then New York–based sustainability consultant JD Capuano worked with a well-known online community platform to map out the environmental impact of their business. The goal was to define and quantify the impact, which included estimating the carbon footprint related to running its website—how much carbon was being released into the atmosphere by its servers' energy use—and devise a strategy for reducing that impact.

Because this company's primary source of income came from its online platform, it was important for it to include data centers and website performance analysis as part of the process. "Our first step was to interview the head of their data center to understand everything they were doing," JD said. The company colocated its servers in a few data centers, meaning it shared server rack space with other companies. It also had a number of geographically distributed servers on multiple content delivery networks (CDNs). A CDN caches redundant copies of content on servers in different geographical areas. When users request that content, it is served from the location closest to them, resulting in speedier delivery and less data transmission.

This was helpful in terms of efficiency, but meant that JD had to spend a lot of time collecting data from disparate sources. "The hardest part was getting some form of usable data from the colocated data center host. It required some negotiation. JD then established a regular feed of this information as it came in to his client's IT team. "The CDNs, on the other hand, were transparent," he said. "One of them even calculated our client's carbon footprint and emailed it over."

JD combined CDN emissions and data center information that was adjusted with partial feeds from power strips that recorded energy the IT Team installed on some racks in the collocated data centers. The power strip feeds let his team quality control the extrapolation method on the provider's data to estimate a unit emission figure that made sense for the organization.

It is important to note that website and data center analysis were part of a larger sustainability audit that JD did for the entire organization. This included measuring and diagnosing its energy, emissions, water, waste, and purchasing impacts. The emissions included its offices, data centers/CDNs, business travel, employee commuting, and shipping.

Combining the overall data culled from the organization's efficiency efforts with that of the online components—website and data centers—enabled JD to make overall recommendations for emissions reductions.

Considering this was the first iteration of its platform's carbon footprint, JD decided to exclude *embodied* energy—the sum of all energy required to produce goods or services—of the equipment, and the energy required at end-of-life for disposal. Separate from GHG emissions, JD devised a process for measuring waste that included e-waste—electronics hardware and other devices—collected in its offices and colocation spaces. He also ensured that they were using a recycler certified by e-Stewards, an organization that provides globally responsible electronics recycling. Chapter 3 includes more information about e-waste.

Sustainable in Digital?

To give you an example of how otherwise innovative and sustainable companies can easily overlook the impact of their digital properties, neither Nascent Objects nor Ecovative Design appear to power their websites with renewable energy. Both sites have room for improvement in performance optimization, as well.

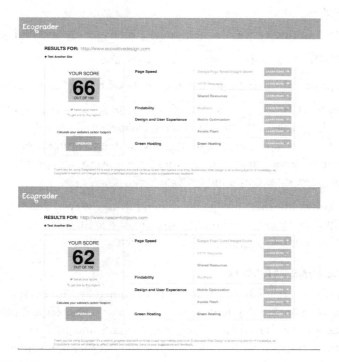

FIGURE 1-12.
Both Nascent Objects and Ecovative Designs have made public commitments to sustainability (hopefully, their websites will soon follow)

That said, these startups are just getting off the ground, so the small amount of emissions their websites produce is not likely a high priority item for either company. They have many other fish to fry in this early stage of their businesses. However, one hopes that as these companies grow and scale, they will look at the impact of their digital products and services alongside the physical ones. The most sustainable startups will be those that bake sustainable thinking into their business model from the beginning—and that includes considering digital products and services.

Sustainability and the Internet

Now that we have covered a brief overview of how sustainability principles are used in business, how can we apply these practices to the Internet? As mentioned previously in this chapter, when it comes to digital considerations, many life cycle assessments stop at employee workstations and data centers. If we look at where energy is used throughout a web application's life cycle, however, we can see that energy is needed in three key areas:

- The creation, testing, launch, and maintenance of these applications
- The hosting and serving of these applications
- The download and interaction of users with these applications

Clearly, making data centers and workstations more energy efficient plays a huge role in the aforementioned areas, but it's not the complete picture. If we want to calculate the environmental impact of a digital product or service during its entire lifetime, we must apply the same LCA process to its entire life cycle, as well.

Let's look at Internet supply and demand from a very high level. Then, we will dive deeper into how you can create a virtual life cycle assessment for your own digital product or service.

THE JEVONS PARADOX

The Jevons paradox is an economic term coined by 19th century economist William Stanley Jevons to explain how technological progress increases efficiency and also increases demand, and subsequently consumption due to what he called a rebound effect—when something is cheap and convenient, more people want it. Jevons was initially talking about coal use. He observed that when technological advancements increased efficiency and reduced prices, it also drove demand for coal and increased consumption across many industries.

Though Jevons was originally talking about energy, his paradox can be applied to almost any resource and is especially relevant to Internet use. In most examples—such as automobiles or lighting—a convenience factor such as price, availability, location, and so on led to widespread adoption, and in some cases innovation. In the context of this book, when more efficiency leads to more consumption, we get 3+ billion

Internet users taking advantage of cheap broadband,[23] inexpensive netbooks, and of course the now ubiquitous smartphone. This number is growing exponentially with some people predicting that the entire world will be online by 2020.[24] Widespread adoption always has impact. The choices we make to buy a cheaper soda, drive a roomier car, or get faster access to the Internet have consequences when they're amplified to include even *half* the world's people. When you consider the amount of virtualization—migrating offline processes online—that global businesses are currently undertaking, the impact is significant. Though these migrations will themselves ostensibly reduce CO_2e emissions, they also have an effect of their own, which is often overlooked.

IOT

Similarly, the IoT will see 50 billion connected devices by 2020, or about 7 devices per person on the planet, according to a popular report by DHL and Cisco.[25] Gartner put the number closer to 20 billion devices.[26] Whether it's a car self-diagnosing faulty brakes or a chip in your pet's collar that texts you Fluffy's location when she has run away, anything that can be assigned an IP address and transmit or receive data is a "thing" in IoT parlance.

The consumer-side promise of the IoT is that its smart, self-aware nature will help people make more responsible choices in real time. On the business side, industrial automation—in everything from traffic lights to buildings—while less visible, will save significant amounts of electricity. Remote monitoring will also save on transportation costs and their accompanying emissions.

"If we can put computing into goods so they can self-describe, and provide access on how to safely dispose of them—where 'dispose of' means return for recycling, or remanufacture, or reuse," says Chris Adams

---

23 Internet Live Stats, "Internet Users". (http://www.internetlivestats.com/internet-users)
24 Chris Greenhough, "Eric Schmidt Predicts Entire World Will Be Online By 2020", Inquisitr, April 15, 2013. (http://www.inquisitr.com/618893/eric-schmidt-predicts-entire-world-will-be-online-by-2020)
25 DHL and Cisco Trend Report 2015, "Internet of Things in Logistics". (http://www.dhl.com/content/dam/Local_Images/go/New_aboutus/innovation/DHLTrendReport_Internet_of_things.pdf)
26 The Register, "Gartner: 20 billion things on the Internet by 2020". (http://www.theregister.co.uk/2015/11/11/gartner_20_billion_things_on_the_internet_by_2020)

from Product Science, a London-based digital agency that works primarily with organizations working on social or environmental problems as part of their business, "then we have a chance to close a lot of extremely wasteful flows of resources."

This is all great for the environment. There are, however, some significant concerns as well:

- Manufacturing all these often disposable devices expends large amounts of energy and uses potentially hazardous raw materials or conflict minerals.
- It is very difficult to gauge the overall footprint of a device because often its components are created in many different places.
- Whereas some, some devices, such as the Nest thermostat (discussed in more detail in a moment), use features and UX to save energy, others, such as fitness monitors or home automation systems, can have much heavier energy footprints.
- IoT devices will often replace older devices, which need to be disposed of. Also, the sometimes disposable/replaceable nature of these devices—their "planned obsolescence"—means that they might end up in a landfill. Some companies even build in self-destruct functionality for security purposes so that a device *can't* be used after a certain date.[27]
- Each device transmits and receives information to a server somewhere in a data center, which requires power 24/7/365. Few of these data centers are powered by renewable energy.

"This ubiquity is a double-edged sword," Chris says. "Computing is now cheap to the point that chips are disposable, so we need to provide other reasons to make them last beyond them just being expensive to deploy, or possible to biodegrade in some safe fashion."

Though all of these *Things* on the IoT will ostensibly give us better analytics for identifying efficiencies and making more sustainable choices, they also require power sources to continue transmitting and receiving data in real time. As mentioned previously, these devices offer

---

27 Klint Finley, "The Internet of Things Could Drown Our Environment in Gadgets", *Wired*, June 5, 2014. (http://www.wired.com/2014/06/green-iot)

unprecedented possibilities for monitoring and controlling energy use, but current IoT standards are scattered at best and very few of the device manufacturers are collaborating to fix the issue. So, although it's great that we can monitor energy use, the lack of data standards can make for unnecessary transmission of information. This makes for a rather significant sustainability challenge when it comes to electricity use. Networks, too, will need to be very robust to accommodate this rapid growth in data transmission.

If you own seven of these devices, for example, and none of them talk to on another, there is likely great redundancy in data transmission as each device transmits and receives in its own proprietary format. While Apple—with its HomeKit framework—and a few others are making in-roads to developing common hardware and software standards for these devices, it is still very much the Wild West, with many companies working in silos. To develop a more sustainable IoT, companies must work together to devise common standards that will streamline data transmission and alleviate redundancies.

There is, however, a great untapped opportunity in IoT to help consumers make more sustainable choices through UX practices when creating interfaces for these devices, as in the case of the Nest thermostat (Figure 1-13).

FIGURE 1-13.
The Nest Thermostat's leaf icon helps users make more sustainable choices

Human behavior is modified more by immediate feedback rather than delayed feedback. Take your electricity bill, for example. You are far less likely to modify electricity use when you receive the bill only every 30 days *after* you've already used the electricity you pay for. The Nest thermostat, on the other hand, not only detects your presence in a room and makes temperature adjustments accordingly to minimize electricity use when you're not around, it also shows a handy leaf icon when you adjust the thermostat to a more energy efficient setting. This instant feedback helps customers make more informed and energy efficient choices.

If immediate feedback isn't enough, Nest also sends a monthly report (Figure 1-14) via email that not only includes information about your own usage, but also compares it to other Nest users in your area, incentivizing cost and energy savings.

Although Nest's use of iconography, real-time feedback, and incentives offer great lessons in creating more sustainable user experiences, there are others, as well. We will cover more on sustainable UX practices in Chapter 5.

Here's how you did:

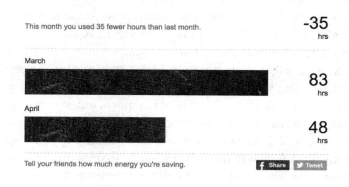

This month you used 35 fewer hours than last month.   **-35** hrs

March   **83** hrs

April   **48** hrs

Tell your friends how much energy you're saving.   [Share] [Tweet]

Why did your energy use change?

We look at a lot of reasons your energy use can change — from weather to Auto-Away — and these are the ones that made the biggest difference this month.

They add up to -38 hours of energy use. The difference of +3 hours was caused by other factors.
Learn more >

-19 hrs — Warmer weather helped you save.

-16 hrs — Your schedule was more efficient this month. *Kudos!*

-3 hrs — April had fewer days than March.

A look at your Leafs:

You get a Leaf when you choose an energy-efficient temperature. This month, the average Nest Thermostat owner in your area earned 15 Leafs. Here's how you did:

In April you earned: **21** Nest Leafs — 1 fewer than March

In April you're in the top: **40%** of Nesters in your area

This year you've earned: **99** Nest Leafs

Let your friends know how many Leafs you earned.   [Share] [Tweet]

FIGURE 1-14.

Nest also sends a helpful monthly email that shows your electricity use as compared to others in your area

RUNAWAY PAGE GROWTH

As global Internet usage has grown, so too have our apps and web pages. As mentioned previously, the average web page size as of this writing is 2.3 MB, more than 24 times the size it was in 2003.[28] When served up on sluggish Internet connections to mobile devices with slower processors, these pages waste time and energy and frustrate the bejeezus out of users. Our love affair with video backgrounds, multi-image carousels, complex social-sharing features, high-res images, rotating banner ads, and other frontend bells and whistles has caused an epidemic of slow, overloaded pages. Meanwhile, many studies have shown that users are unlikely to wait more than a couple seconds for a page to load. There seems to be a bit of cognitive dissonance happening here.

Chapter 6 is devoted entirely to helping you optimize pages for better performance and faster delivery.

VIDEO STREAMING

Let's not forget one of the most bandwidth-intensive uses of the Internet to date: streaming media. Consider these statistics:

- With more than 1 billion users as of this writing, YouTube alone streams hundreds of millions of hours of content each day. 300 hours of video are uploaded every *minute*. Still, according to broadband Internet tracking firm Sandvine, YouTube made up about 18% of overall Internet downstream traffic at the end of 2015.[29]

- Netflix use, on the other hand, made up about 37% of all Internet downstream traffic.[30]

- Amazon video comprises slightly more than 3%.

---

28 Website Optimization, "Average Web Page Breaks 1600K". (*http://www.websiteoptimization.com/speed/tweak/average-web-page*)

29 Emil Protalinski, "Streaming Services Now Account for Over 70% of Peak Traffic in North America, Netflix Dominates with 37%", VentureBeat, December 7, 2015. (*http://venturebeat.com/2015/12/07/streaming-services-now-account-for-over-70-of-peak-traffic-in-north-america-netflix-dominates-with-37*)

30 Sandvine, "Global Internet Phenomena Report". (*https://www.sandvine.com/trends/global-internet-phenomena*)

In all, video streaming services like YouTube, Netflix, Hulu, Vudu, and others already make up more than 70% of consumer Internet traffic.[31] In 2013, Cisco said streaming video was expected to grow to 76% by 2018 but by current numbers, that estimate might be quite low.[32] One can see why it is important that the electricity used to power these services is generated by renewable sources. Currently, there are big drawbacks to that, which we will cover in more detail in Chapter 3.

Progress is being made on this front. In 2013, according to Google Green, Google's website on its environmental impact, 35% of the company's energy came directly from renewable sources, leaving 65% provided by other nonrenewable resources. Greenpeace's 2015 *Clicking Clean* report, which bills itself as a "a guide to building the green Internet" put that number at 46%, with 15% coming from natural gas, 21% from coal, and 13% from nuclear sources. Google claims to offset its nonrenewable power sources, but as we will see in Chapter 3, renewable energy credits have their own set of drawbacks. Still, with more than $2.5 billion worth of investments in renewable energy projects and a commitment to 100% renewable power by 2025, it is one of the largest corporate investors in renewable energy in the world.[33]

Another example: Netflix and several others such as Vimeo, Vine, and SoundCloud use Amazon Web Services (AWS) to host their content, as do hundreds of other popular apps we all use, like Dropbox, Pinterest, The Huffington Post, Yelp, Reddit, and so on. In fact, a 2012 study by Deepfield noted that one-third of all Internet users will access an AWS-hosted site or app on average of at least once per day.[34]

This can be problematic for sustainability purposes because AWS has consistently come under fire by Greenpeace and other organizations for its lack of transparency about energy use and sources. Though it has made a long-term public goal to be 100% renewable energy powered, the lack of transparency from AWS makes it difficult for the public to

---

31 Emil Protalinski, "Streaming Services". (*http://venturebeat.com/2015/12/07/streaming-services-now-account-for-over-70-of-peak-traffic-in-north-america-netflix-dominates-with-37*)
32 Cisco Visual Networking Index: Forecast and Methodology, 2013– 2018 (2014).
33 Google, "Renewable Energy–Google Green". (*https://www.google.com/green/energy*)
34 Robert McMillan, "Amazon's Secretive Cloud Carries 1 Percent of the Internet", Wired, April 18, 2012. (*http://www.wired.com/2012/04/amazon-cloud*)

ascertain how it will actually accomplish this goal, leading Greenpeace to grade the company with C, D, and F ratings across its 2015 Clicking Clean Scorecard.[35]

With more than 60% of the Fortune 100 having carbon and renewable energy goals in place, the company's lack of transparency will likely become of larger concern to their customers.[36] But they are not the only player for which this has become an issue. As you can see from the chart in Figure 1-15, other companies such as eBay and Digital Realty, which hosts LinkedIn, have a ways to go, too.

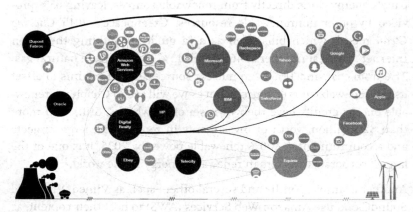

FIGURE 1-15.
This infographic from Greenpeace's 2015 *Clicking Clean* report shows which of the big Internet companies have committed to renewables and which still have a ways to go

When you couple the meteoric growth of streaming video with the lack of transparency and available resources for hosting powered by renewable energy, the problem becomes crystal clear. Again, Jevons paradox in action.

---

[35] http://www.clickclean.org

[36] Ceres, "Power Forward 2.0: How American Companies Are Setting Clean Energy Targets and Capturing Greater Business Value". (http://www.ceres.org/resources/reports/power-forward-2.0-how-american-companies-are-setting-clean-energy-targets-and-capturing-greater-business-value/view)

VIRTUAL REALITY

With the release of devices like the Oculus Rift, Sony's PlayStation VR, and others entering consumer homes for the first time in 2016, virtual reality (VR) has seen a resurgence in popularity that Dr. Jonathan Waldern, a VR pioneer, calls "as big an opportunity as the Internet."[37] Projected revenue by 2020 is expected to be at $5 billion for sales of virtual reality game content, but that's just scratching the surface. Projected 2020 sales for virtual and augmented reality content are expected to be at $150 billion.[38] Although traditional game markets make logical sense, online providers like Netflix, Hulu, and Amazon are also reportedly exploring ways to add VR content to their subscription services. This could put significant pressure on already stressed bandwidth.

FIGURE 1-16.
Virtual reality: an opportunity as big as the Internet

The ultimate goal of virtual reality is to create as much detail in the rendering of a virtual setting that it is indiscernible from the real thing. This includes visual, aural, touch, temperature, and eventually

---

37 Charles Arthur, "The Return of Virtual Reality: 'This Is as Big an Opportunity as the Internet'", *The Guardian*, May 28, 2015. (*https://www.theguardian.com/technology/2015/may/28/jonathan-waldern-return-virtual-reality-as-big-an-opportunity-as-internet*)

38 Janko Roettgers, "Hardware Giants Bet Big on Virtual Reality and a Market That Doesn't Yet Exist", *Variety*, March 22, 2016. (*http://variety.com/2016/digital/news/virtual-reality-oculus-rift-consumers-1201735290*)

olfactory details. According to an early 2016 article in *Forbes*, "humans can process an equivalent of nearly 5.2 gigabits per second of sound and light—200 times what the US Federal Communications Commission predicts to be the future requirement for broadband networks (25 Mbps)."[39] Because it is estimated that the human eye can perceive up to 150 frames per second, "assuming no head or body rotation, the eye can receive 720 million pixels for each of 2 eyes, at 36 bits per pixel for full color and at 60 frames per second: that's 3.1 trillion (tera) bits!"

We're not there yet, of course. And although compression algorithms can reduce some of this, for VR to realize its true potential, the fact remains that exorbitant amounts of bandwidth will be required to transfer huge data payloads to end users.

According to Dr. Markiewicz:

> The game industry, which will have a major role in virtual reality, is not known for efficiency or environmental awareness. The cultural zeitgeist glorifies Hummer computers over Prius rigs, and game design has emphasized ever more elaborate 3D and animation without regard for power consumption. If graphic design has ignored the energy consequences of bad web design, the game industry celebrates its defiance of sustainable norms.

With all this in mind, virtual reality content creators will need to embrace new breakthroughs in performance optimization to maintain a good balance between limited bandwidth, great UX, and optimized performance without latency.

DATA CENTERS

Finally, no discussion of Internet sustainability would be complete without mentioning the power used by data centers. From those owned by large companies like Google and Facebook to the colocated servers used by web design firms and small startups across the globe, every server on the Internet pulls energy from our aging electricity grid 24/7/365. Nearly all of them have redundant methods in place to serve data from alternative sources if a server goes down and to guard against

---

39 Bo Begole, "Why The Internet Pipes Will Burst When Virtual Reality Takes Off", *Forbes*, February 9, 2016. (http://www.forbes.com/sites/valleyvoices/2016/02/09/why-the-internet-pipes-will-burst-if-virtual-reality-takes-off/#774f287064e8)

power failures, so these redundancies also require power in addition to that used by their live servers. They also typically have banks of generators—which run on diesel fuel—as backup power sources.

This is important because according to a 2013 post in the *New York Times*, digital warehouses use about 30 billion watts of electricity, about the same output as 30 nuclear power plants.[40] A single data center can use as much electricity as a small town (Figure 1-17). According to that same article, on average, data centers only use 6 to 12% of the electricity powering their servers for computations. The rest is used to cool servers down, keep them idling, and to guard against surges that could crash their systems. In other words, it is possible that data centers can waste around 90% of the electricity they pull off the grid.

FIGURE 1-17.
A single data center can use as much electricity as a small town and many also use diesel-fueled backup generators to maintain uptime during outages

---

40 James Glanz, "Power, Pollution and the Internet", *New York Times*, September 22, 2012. (http://www.nytimes.com/2012/09/23/technology/data-centers-waste-vast-amounts-of-energy-belying-industry-image.html)

In countries like the United States, only about 13% of the nation's energy comes from renewable sources, according to the US Energy Information Administration.[41] This includes hydroelectric power, which some environmentalists don't consider a true renewable source because of the potential damage it can have on natural habitats, but doesn't include nuclear power (another 20%), which has environmental issues of its own.

When one considers this alongside the 30 billion watts of energy used by data centers, 90% of which is wasted, one can quickly see a picture emerging of an industry poised to be one of the planet's biggest polluters.

Moving to renewables

Although some of the larger Internet companies like Apple, Google, and Facebook have made huge strides in powering their centers with renewable energy, as of this writing, others, like Microsoft and AWS, still have a ways to go yet. Many cloud providers, like Heroku, as an example, build on top of AWS's existing infrastructure. In late 2015, Rackspace, which struggled as a competitor to AWS in cloud-based services, followed Heroku and others' path by offering managed cloud services on AWS.[42]

Similarly, in the world of colocation and web hosting providers, some companies have made strides toward powering their centers with renewables, but many are completely opaque about where their energy comes from, including providers that tout their "green" credentials. This is tremendously important to conscientious designers and developers who want to build environmentally friendly digital products and services. Where we host our apps matters.

We will cover data centers, hosting providers, and renewable energy in more detail in Chapter 3.

---

41 US Energy Information Administration, "How Much US Electricity Is Generated from Renewable Energy?". (http://www.eia.gov/energy_in_brief/article/renewable_electricity.cfm)

42 Yevgeniy Sverdlik, "Rackspace to Provide Managed AWS Services Before Year's End", Data Center Knowledge, August 12, 2015. (http://www.datacenterknowledge.com/archives/2015/08/12/rackspace-provide-managed-aws-services-years-end)

Virtual LCAs

Now that we have a better understanding how and where energy is used to power the Internet and a working knowledge of sustainability principles in business, let's apply the LCA to digital products and services. Remember that list of processes we discussed earlier in the section on inventory analysis? Here is where that specifically applies to digital products and services.

In an article published on Creative Bloq (formerly *.Net* magazine), Dr. Pete Markiewicz proposed a framework for equating virtual LCAs to their physical counterparts.[43] Components include those listed in Table 1-1.

TABLE 1-1. This chart from Pete Markiewicz compares standard life cycle assessment components to their digital counterparts

LCA	VIRTUAL LCA
Materials	Software and visual assets
Manufacturing	Design and development
Packaging	Uploaded to the Internet
Distribution	Downloaded through the network
Usage	Interaction, user experience, completing tasks
Disposal	Data erased from client

According to Dr. Markiewicz:

> Compared to the physical world, sustainable web design will de-emphasise the manufacturing (read production) stage, making it qualitatively different from print design. Why? Unlike physical products, web pages don't produce waste paper or ink after the page disappears, only heat from the electronics. For this reason, the cost of web page manufacture will be small compared to the cost of ongoing use. On the other hand, the longer a web page is viewed, the more bits it burns, so efficient use based on good user experience will be very important.

Let's take a look at each of these virtual components individually.

---

[43] Pete Markiewicz, "Save the Planet Through Sustainable Web Design", Creative Bloq, August 17, 2012. (http://www.creativebloq.com/inspiration/save-planet-through-sustainable-web-design-8126147)

SOFTWARE AND VISUAL ASSETS

Rather than asking how much energy a manufacturing process uses, let's look at the software and visual assets needed to build our digital products and services. If you use cloud-based services, are they hosted on servers powered by renewable energy? If not, do you tend to leave applications open and running in the background on your computer? Many of these applications regularly use network connections to check for updates or provide usage information to the company that created them. When left open, they also use RAM and processing power, which wastes energy.

It can be next to impossible to decipher how energy efficient the creation process of a piece of software is. Did they use Lean Startup principles, creating only what was necessary to validate their product? Or did they load the software package up with features that are rarely used? (I'm looking at you Adobe Photoshop and Microsoft Word.) Are the software company's offices powered by renewables? Is the equipment energy efficient? The same questions apply to visual assets used during the creation process. It can be challenging to answer these questions, but if we are equating the manufacturing process of a physical product to that of a virtual one, they are worth considering in the name of reducing wasted and emissions.

DESIGN AND DEVELOPMENT

It is in the design and development of a product where you can make many efficiency gains and energy savings. Strategies like progressive enhancement—an approach that uses feature layering to allow anyone access to basic content and functionality while also providing enhanced features to those with more advanced browsers and/or greater bandwidth—and mobile-first makes digital products and services accessible to more people. Web standards like HTML5 and CSS 3 as well as emerging open source frameworks for virtual reality like WebVR and OSVR will allow for more efficient delivery of content to users across devices. Similarly, accessibility standards for users with disabilities who might need to access your content with screen readers or other assistive technologies will mean that your content can be viewed across a wider array of platforms and devices.

When running an inventory analysis on a digital product's design and development, some questions to ask include the following:

- Does the product use outdated or nonstandard technology that make hardware work harder or requires plug-ins or other additional resources to run?
- Is it optimized for mobile devices?
- Are images, scripts, and other assets compressed or otherwise "minified" for fastest possible delivery?

SERVER UPLOADS, NETWORK DOWNLOADS

Our digital products and services might not require packaging or cargo ships to transmit them from place to place, but there is still energy needed to get these applications to their virtual destinies. The US Energy Information Administration estimates that about 6% of electricity used in the United States is lost in transmission and distribution.[44] Typical modernized countries have similar rates between 6 to 9%.[45] In developing countries, it's another story. In 2000, India lost around 30% of its electricity in transmission, but has made improvements and is currently at about 18%. This is due to the simple fact that by running electrical energy through wires some of it eventually becomes heat.

Designers and developers can create applications that are more efficient and served from computers powered by renewable energy to make this less of a problem. Use of CDNs, by which pages are served from the server closest to the person requesting them can help minimize this, as well.

Thus, when running an inventory analysis on a digital product or service, it is important to ask questions about how much energy is lost in transmission *as well as* that which is used by infrastructure, bandwidth, and frontend use. Are there ways to reduce those amounts?

---

44 US Energy Information Administration, "Frequently Asked Questions". (*http://www.eia.gov/tools/faqs/faq.cfm?id=105&t=3*)

45 Robert Wilson, "How Much Electricity Is Lost in Transmission?", Carbon Counter, February 15, 2015. (*https://carboncounter.wordpress.com/2015/02/15/how-much-electricity-is-lost-in-transmission*)

INTERACTION

All digital products and services require energy to run. Interaction represents the most profound way they differ from their physical counterparts, as noted by Dr. Markiewicz's quote earlier. Question is, how *much* energy do they use? How can one answer this for a virtual LCA inventory? Here are some questions to ask:

- How much traffic does my site or app get?
- How much of that traffic comprises computers versus mobile devices?
- How much energy do each of those devices use per second? Per minute?
- How much data is uploaded/downloaded per device per minute?
- What is the average amount of time spent (per device) using my digital product or service?

You can find answers to many of these questions via Google Analytics or other metrics measurement applications.

DISPOSAL OF DATA

In the inventory analysis for a physical product, one considers how much waste is created and energy used when a product is discarded. Often the focus is on recycling disposed materials. With a virtual product, recycling isn't an issue, though churn can leave residual data for closed accounts in databases, increasing an application's size and hence footprint. On the frontend, power is used to delete applications or documents, though that is minimal in comparison to usage and interaction. Also, many online systems, like Git, for example, back up data rather than deleting it. So, if you store thousands of backups, even if only changes are stored, does that increase electricity use and hence environmental impact while also decreasing efficiency? Although disk storage might be energy efficient, backup management might be inefficient. Also, as our applications grow in size and complexity, this can become more of an issue.

Some questions to ask include the following:

- What processor resources are required to delete my application?
- Does it leave residual information on the user's device?
- How many unused accounts are in the database?

Code obsolescence

Many digital products and services we rely on every day are built on a foundation of obsolete code and outdated proprietary systems, hacked together with the digital equivalent of dental floss and chewing gum. Airline reservations and stock trading are just two examples of large systems our society depends upon to function but which sit upon unstable and outdated code. In July 2015, both the NYSE and United Airlines' systems went down on the same day (as did the *Wall Street Journal*, for that matter). Outdated code not only leads to potential security and performance issues, but it also wastes energy and puts your users' data at risk.

Investing in updates

The importance of keeping hardware/software up to date cannot be stressed enough. Band-aid fixes without the equivalent accompanying architecture planning, budgeting, and so on add more vulnerability to your digital product or service. The duct-tape method of coding new features is unfortunately more common than one might think and is decidedly not a very sustainable approach. In a Medium post about this topic, professor and software developer Zeynep Tufekci, states:[46]

> A lot of new code is written very very fast, because that's what the intersection of the current wave of software development (and the angel investor/venture capital model of funding) in Silicon Valley compels people to do...Software engineers do what they can, as fast as they can. Essentially, there is a lot of equivalent of "duct-tape" in the code, holding things together. If done right, that code will eventually be fixed, commented (explanations written up so the next programmer knows what the heck is up) and ported to systems built for the right scale—before there is a crisis. How often does that get done? I wager that many wait to see if the system comes crashing down, necessitating the fix. By then, you are probably too big to go down for too long, so there's the temptation for more duct tape.

---

46 Zeynep Tufekci, "Why the Great Glitch of July 8th Should Scare You", The Medium, July 8, 2015. (*https://medium.com/message/why-the-great-glitch-of-july-8th-should-scare-you-b791002fff03*)

Also, if you build an application that relies on someone else's data, keep in mind that every time they update their application programming interface (API), you might likely need to update your application, as well. Sometimes, this will result in features users love no longer being available. A digital product or service will see its life cycle cut short if the software isn't fast, reliable, and up to date.

When running an inventory analysis on the end stages of your digital product or service, ask the following questions:

- How will you keep your application updated, your code lean, and your users' data secure over time?
- How will you add new features that users want while still staying efficient?
- When the time does come to finally put your app to rest, how will you do so with as little waste as possible?
- What tools can you add to your workflow (e.g., testing modules, real-time performance analytics, hack-ability tests) to improve code quality?
- What project management strategies can you implement to prevent waste of coding time, improve bug detection, and stress-test applications during Agile sprints? (More on Agile in Chapter 3.)

Hardware Disposal

FIGURE 1-18.
The US produces enough annual e-waste to fill 1.15 million 18-wheel trucks. Put end-to-end, those trucks would stretch nearly halfway around the world.

Outdated hardware can have a significant impact on performance, stability, and security. Sometimes, it needs to be replaced and should be done so in a manner in which the hardware components can be recycled wherever possible. As of May 2015, up to 90% of the world's electronic waste is illegally traded or dumped each year, according to the United Nations.[1] This hardware contains toxic chemicals like mercury, lead, cadmium, and arsenic, which can cause cancer, reproductive disorders, and other health problems. The United States is the world's largest producer of e-waste, more than one million tons ahead of China. How much waste are we talking about? 41.8 million metric tons in 2014 alone, enough to fill 1.15 million 18-wheel trucks. With the average truck trailer being about 53 feet long, lined end-to-end you could reach nearly halfway around the world with all the e-waste the United States produces in a single year.

---

1 America Upcycles, "About e-Waste". (*http://www.americaupcycles.com/#!e-waste/xx7tv*)

Conclusion

The statistics shared in this chapter are meant to give context. The Internet provides great social and environmental benefits by allowing easy access to data and information that was once locked up in processes with far greater environmental impact. (After all, when was the last time you purchased a CD at a record store?) This easy access has empowered nations, enabled communications where previously there were none, and given people in some societies their only access to important information like medical treatment or education.

The solution, of course, is not to use the Internet less. It is to make the system more efficient, power it with renewable energy, and take advantage of it to solve large, global problems like inequality or the climate crisis.

Action Items

Want to explore the concepts outlined in this chapter even further? Try these things:

- Create a list of areas where you could have made potentially more sustainable choices on the last project you worked on.
- Use the framework in this chapter to run a virtual LCA of a website or application you have either built yourself or use regularly. Identify areas for improvement. Where are there gaps?
- Ask your hosting provider what its policy is for using renewable energy to power its servers.

After you have identified areas for improvement, you can use the framework described in the following chapters to begin making your digital products and services more sustainable.

[2]

A Sustainable Web Design Primer

What You Will Learn in This Chapter
In this chapter, we will cover:

- Why standards for Internet sustainability are important.
- The individual components of a framework for creating more sustainable digital products and services.
- Potential barriers to widespread adoption of these standards.
- Workarounds for building awareness and decreasing environmental impact.

Sustainable Web Design
In December 2015, leaders from 195 countries came together at the United Nations Conference on Climate Change, adopting the first ever universal climate accord for reducing greenhouse gas (GHG) emissions in an effort to slow or halt global warming. After much deliberation, they agreed to the following:

- A goal for reducing the global temperature rise to well below 2° C above preindustrial levels.
- Participating countries will work to monitor, report, and curb their GHG emissions.
- Countries will invest hundreds of billions of dollars in climate-related financing toward curbing emissions and expanding cleaner, renewable energy sources.

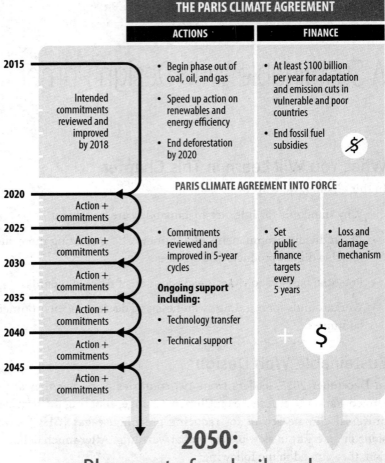

FIGURE 2-1.
At COP21 in Paris, leaders from 195 countries decided on a list of actions and commitments to be free of coal, oil, and gas with 100% renewable energy by 2050

On Earth Day 2016, 175 countries committed to signing the Paris Agreement on Climate Change at the United Nations headquarters in New York.[1] United States Secretary of State John Kerry called the agreement the "strongest, most ambitious climate pact ever negotiated."[2]

The Paris event was the largest ever gathering of world leaders negotiating to reach an agreed-upon deal around climate change. Reporters from *The Guardian* called it "the world's greatest diplomatic success"[3] while Komi Naidoo, executive director of Greenpeace International, noted that, "Today the human race has joined a common cause...The deal alone won't dig us out of the hole that we're in, but it makes the sides less steep."

Although this is a monumental and historic step in the right direction, it remains to be seen how, specifically, the accord will be executed and what long-term effects will come from the subsequent efforts. Many have noted that these efforts are potentially too little, too late, but all agreed that COP21 marked significant progress in the fight for our planet's future.

"Any long-term solution will require profound changes in how we generate energy," according to Josh Katz and Jennifer Daniel in their *New York Times* piece "What You Can Do About Climate Change."[4] "At the same time, there are everyday things that you can do to reduce your personal contribution to a warming planet."

With the Internet playing an increasingly important role in the daily lives of billions of people, there is a moral imperative for our community to address these issues and make creating more sustainable digital products and services something that's easy for everyone to understand and implement. Frontend energy consumption of any new digital product or service is typically determined during the design phase, so keeping design teams informed of these issues can make a significant

---

1 United Nations, "List of Parties that Signed the Paris Agreement on 22 April". (*http://www.un.org/sustainabledevelopment/blog/2016/04/parisagreementsingatures*)

2 Michele Gorman, "150 Countries Will Sign Paris Climate Change Agreement on Earth Day", *Newsweek*, April 22, 2016. (*http://www.newsweek.com/countries-sign-paris-climate-change-agreement-earth-day-451443*)

3 Fiona Harvey, "Paris Climate Change Agreement: The World's Greatest Diplomatic Success", *The Guardian*, December 14, 2015. (*http://www.theguardian.com/environment/2015/dec/13/paris-climate-deal-cop-diplomacy-developing-united-nations*)

4 Josh Katz and Jennifer Daniel, "What You Can Do About Climate Change", *New York Times - The Upshot*, December 2, 2015. (*http://www.nytimes.com/interactive/2015/12/03/upshot/what-you-can-do-about-climate-change.html*)

difference. Per Katz and Daniel's article, there are numerous everyday things designers and developers can do to create better solutions. This chapter proposes a framework for creating and maintaining more sustainable digital products and services to address the issues outlined in Chapter 1. The individual topics within this framework are addressed in detail throughout the rest of the book.

WEB SUSTAINABILITY STANDARDS?

Many industries already have well-established standards for sustainability. LEED, Passiv Haus, Living Building Challenge, Net Zero, and other ratings systems, for example, can help builders find the best route toward creating an environmentally friendly, emissions-neutral structure. On the Internet, although we have professional standards set forth by the World Wide Web Consortium (W3C), none of these to date take energy consumption or GHG emissions into consideration whatsoever. The best we have currently is an open W3C community group focused on sustainable web design (which you should join).

We need any easy way for web teams to better understand how they can build more sustainable web solutions. Should we get designers and developers to embrace standards for sustainability so that in several years the Internet reaches where the building industry is now?

The folks at Future Friendly, a collective of Internet authors, designers, and luminaries, don't think standards are the way to go.[5] Things are changing too quickly. As written on the front page of their website:

> Our existing standards, workflows, and infrastructure won't hold up. Today's onslaught of devices is already pushing them to the breaking point. They can't withstand what's ahead. Proprietary solutions will dominate at first. Innovation necessarily precedes standardization. Technologists will scramble to these solutions before realizing (yet again) that a standardized platform is needed to maintain sanity. The standards process will be painfully slow. We will struggle with (and eventually agree upon) appropriate standards. During this period, the web will fall even further behind proprietary solutions.

---

5 *http://futurefriendlyweb.com/index.html*

But, according to the Future Friendly crew, there is hope, that web designers should embrace the following principles:

- Acknowledge and embrace unpredictability.
- Think and behave in a future-friendly way.
- Help others do the same.

What could be more future-friendly than acting on climate change? To think and behave this way, Future Friendly advises that we be less stringent on specific methods, technologies, or workflows in order to keep progress and innovation moving forward:[6]

> We can't be all things on all devices. To manage in a world of ever-increasing device complexity, we need to focus on what matters most to our customers and businesses. Not by building lowest common-denominator solutions but by creating meaningful content and services. People are also increasingly tired of excessive noise and finding ways to simplify things for themselves. Focus your service before your customers and increasing diversity do it for you.

FIGURE 2-2.
Standards are slow but technology moves fast

This is good advice. By focusing on the most meaningful content and services, we can provide more relevant experiences, which can be interpreted as a more sustainable approach. But is it the best way to also not compromise the needs of the future, as sustainability principles dictate? The most sustainable solution is one that quickly provides the most pertinent content across the widest array of devices, purportedly as far into the future as possible. Thus, we find ourselves in a common technology quandary: risk alienating users on older devices with a subpar

---

6   *http://futurefriendlyweb.com/thinking.html*

experience or create so many contingencies and workarounds that our applications suffer from performance bloat. Eventually web standards teams might figure a way to remedy this for good, but for now, tactics such as progressive enhancement and a content-first approach (which we'll explore further in Chapters 4 and 5) can help.

Mindset shift

Although this book includes many of the aforementioned methods and workflows, it is important to reiterate that creating more sustainable digital products and services requires a mindset shift, as we covered in Chapter 1, more than it does embracing a particular practice or technique. We're still going to cover a lot of techniques, but we'll do so with the understanding that there are many ways to reach the goal of creating faster, more efficient user experiences powered by renewable energy. It's up to each of us to chart our own course based on our unique needs, rooted in some easy-to-understand principles.

More sustainable solutions are also more efficient solutions. More efficient solutions typically yield greater success. As Nathan Shedroff says in his book, *Design Is the Problem*, "Sustainable organizations are often more successful when they pay attention to the details of waste and impacts, allowing them to function more cleanly, increase profit margins, and differentiate themselves from other organizations."

With that in mind, here's where we're at right now:

- Current web design and development standards don't take energy use or environmental concerns into account whatsoever. Most people don't even know that this is a thing.

- There's no framework for web teams to have meaningful dialogues, either among themselves or with stakeholders, about the potential environmental impact of design and development decisions.

- A significant portion of the Internet's total footprint—some say up to 40%—occurs on the frontend, the part created by designers.[7]

- Although statistics exist on the overall environmental impact of the web, widely accepted methodologies for calculating the impact of individual products or services don't exist, making it tough for product owners to truly understand why this matters.

---

7 James Christie, "Sustainable Web Design", A List Apart, September 24, 2013. (*http://alistapart.com/article/sustainable-web-design*)

- Because much of the Internet is woefully unsustainable, these tactics offer huge opportunities for designers and developers to grow their businesses while making a real difference.

Standards or not, for the aforementioned reasons it makes sense to devise a framework that covers the most common areas where energy use can be minimized while keeping users happy and performance optimized. So here we go.

SUSTAINABLE WEB DESIGN: A FRAMEWORK

Sustainable web design is based on standard environmental conservation principles that can be applied to the life cycle of a digital product, service, or any type of online media. These principles maximize the efficiency of the web applications and media we create, reducing their carbon footprint and decreasing environmental impact. Sustainable web design principles generally focus on reducing electricity use but also cover the inclusion of "green" ingredients, such as clean energy–powered web hosting, for example. These principles also help users make more sustainable choices and meet their needs quickly.

FIGURE 2-3.
A framework for building more sustainable digital products and services: renewable energy, design and user experience, performance optimization, and findability of content

The sustainable web design framework in this book includes four primary categories:

- More sustainable components, like green hosting
- Findability, content strategy, and SEO
- Design and user experience
- Web performance optimization (WPO)

There has already been much written in each of these categories, especially the last three. Many volumes exist that dive deeply into each. Our purpose is to offer a set of guidelines grounded in these existing principles while viewing each through the lens of sustainability.

If embraced, these principles offer the following benefits:

Improved performance
    Lean, fast-loading pages.

Improved usability
    Clear navigation and messaging.

Improved search engine results
    Easily crawled websites that drive traffic and deliver answers quickly.

Improved accessibility
    Relevant content delivered quickly across platforms and devices.

Decreased environmental impact
    Solutions powered by renewable energy.

Let's briefly define each of the four principles here. The chapters that follow will break them down in further detail.

More sustainable components

Hosting your digital product or service on servers powered by 100% renewable energy is the most important thing you can do to create more sustainable online solutions. But it's not the only thing. Many components that go into creating digital products and services can be made more sustainable to create output that meets better environmental standards. Some, such as environmentally friendly workspaces and green mission statements, provide goals and context for sustainable design efforts. Others, like energy-efficient web frameworks, Agile workflows, and standards-based development, directly affect the digital product or service itself. You can't bake a great cake without great ingredients. The same goes for more sustainable digital products and services. How these components fit into the larger mix is what we will explore in Chapter 3.

Here are some questions that typically arise:

- Are your digital products and services hosted by providers that use renewable energy to power their servers?

- Do your providers take advantage of content delivery networks (CDNs), shared libraries, and other external components that can make the end product or service more efficient?

- Does the office environment in which your digital products and services are created support and encourage more sustainable choices?

- Do the people creating these solutions clearly understand the environmental mission of the product, service, and company creating them?

- Are the workflows in which these solutions are devised more sustainable in and of themselves? Do they create less waste?

- Are there other "green" components that could go into creating more sustainable solutions?

FIGURE 2-4.
Green hosting is just one of many components that go into creating more sustainable digital products and services

FIGURE2-5.
Environmentally friendly office spaces include LED lighting, recycling, natural light, and in this case, a living plant wall

Findability and content strategy

Easy-to-find content uses fewer resources. Using good search engine optimization (SEO) practices to improve search engine ranking isn't just a good marketing practice; it is a potentially more sustainable choice and better for users, as well. Similarly, adding search to your own site helps users find things more quickly. These are also more sustainable practices because they use less frontend energy. In other words, better SEO/search on site = content found more quickly = less energy used = more sustainable.

But it doesn't end at search results. After it is found, your content should clearly and quickly serve its intended purpose and help users make more sustainable choices, too—like offering more sustainable shipping or highlighting more ethically produced products first. This requires good content strategy practices. If you're a content creator, the choices you make to promote your content have implications, as well. Chapter 4 discusses this in detail.

Some questions that typically arise here include the following:

- Can users easily find your content through search engines?
- What can you do to improve content to make it more findable?
- Is the content's purpose clear and engaging from the start?
- Does the content clearly provide value and answer questions users will have on a specific topic? Is it helpful?
- Does the content help users make more sustainable choices in terms of shipping, printing, and so on?
- Can users find content easily within the product or service through use of internal search?
- Is there a feedback mechanism corresponding with your content to make it more usable?
- Is there a direct correlation between content promotion and business needs? How do you measure success?

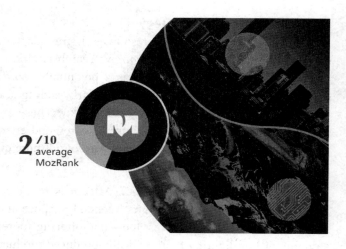

FIGURE 2-6.
Find it fast, use less energy. The average MozRank of nearly 70,000 sites crawled by web sustainability tool Ecograder was only 2 out of 10—big room for improvement here.

Design and user experience

Design and user experience (UX) are where the seeds of web sustainability are sown. Products and services that provide a streamlined yet enjoyable experience—putting the right things in front of users at precisely the moment needed and nothing more—are more efficient and more sustainable. UX designers are in a unique position to create tools with sustainability at their heart by streamlining user workflows, minimizing information overload, and removing potential distractions that keep users from accomplishing tasks they set out to do. In the case of sustainable UX, people-friendly is also more planet-friendly. We explore this further in Chapter 5.

Here are some questions that typically arise:

- Does the interface at all get in the way of users accomplishing tasks quickly and efficiently?
- Is the navigation easy to understand?
- Are design patterns based on commonly accepted standards?
- Can users have a relevant experience no matter their device or browser?

- Are you avoiding "dark patterns" in UX that trick users into increasing virality while undermining integrity? Are you practicing ethical UX?
- Does the experience avoid proprietary formats and plug-ins?
- Does the site use progressive enhancement, mobile-first, HTML5, CSS 3, and so on?

FIGURE 2-7.
Submit it fast. Can users accomplish tasks quickly and easily? Of nearly 70,000 sites crawled by Ecograder less than one-quarter of them were optimized for mobile devices.

Web performance optimization (WPO)

Digital products and services that are speedy and reliable are also more sustainable because they typically use fewer resources to drive more meaningful experiences for users. In fact, studies show that the majority of users might leave your site for a competitor's if it doesn't load within two seconds. Implementing WPO on all the elements of your application will dramatically improve speed and use less energy on both the frontend and the backend. But speed alone doesn't guarantee reliability. Often, more reliability can also mean more overhead. Chapter 6 provides a general overview of performance optimization tactics and discusses how to strike a balance between speed and reliability when creating more sustainable digital products and services.

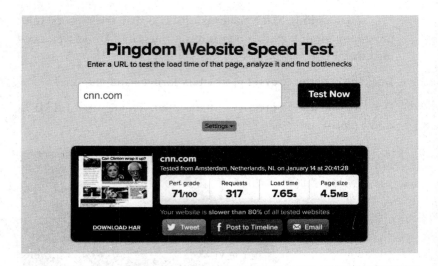

FIGURE 2-8.
Make it lean, serve it fast—tools like those from Pingdom can help you better understand how to improve performance

Some questions that typically arise here include:

- How fast *should* my website or application load?
- How do I balance performance goals with design decisions and client requests?
- Does prioritizing optimization in any way undermine UX or reliability?

- Is the product or service standards-compliant? Does it work for people with disabilities or those on older mobile devices? Does it provide the best possible performance with the least amount of resources across devices and platforms?
- When should I test performance? When should I begin optimizing?
- Will sharing widgets or blog comments slow down my site? How can I fix that?
- How can I keep a commitment to performance and reliability without blowing my budget?

POTENTIAL BARRIERS AND WORKAROUNDS

Although these principles are encouraging, there are some pretty significant barriers to reaching the vision of an Internet that is powered by renewable energy and where efficiency and users reign supreme.

Awareness

When I present this information before a group—be that audience composed of clients, environmentalists, web developers, marketers, or anyone else—I am often greeted by bewilderment. Most folks haven't considered that this is a thing. Sure, some are skeptics, but the majority of audience members simply haven't considered the concept of Internet sustainability before, not because they didn't care but because the concept never occurred to them. When pressed further, some respond that they always considered the Internet to be a greener medium because it replaces paper.

At Mightybytes, we have some clients who are concerned about things like green hosting and want to embrace sustainable web design concepts, but it's usually after we explain the situation to them. The majority don't know enough to care...or just simply don't think it should be a priority. So, clearly awareness of this issue is probably its biggest hurdle.

Adoption

As mentioned at the beginning of this chapter, the physical construction and architecture industries have standards for more sustainable building. They aren't perfect, and as with many things that aspire to higher standards, are always evolving. But they exist. Our industry isn't there yet.

To reach wider adoption of more sustainable standards for creating digital products and services, the folks who define overall web standards should consider energy efficiency alongside other metrics used for defining success. Sure, standards move slow, as we noted earlier, and more future-friendly thinking requires that we embrace unpredictability. But energy efficiency isn't a new concept. It's been around for a very long time. It's just rarely applied to web design. Most folks have historically focused on servers.

Speaking of servers, the web hosting industry is unregulated and greenwashing is as much a reality there as in any other unsupervised industry. Setting and enforcing easily comprehendible standards for web hosts can help people make more educated choices in how or where they host their applications.

Finally, schools churn out thousands of graduates from interactive media programs around the globe every year. Very few of those students enter the workforce aware of the Internet's environmental impact even though they represent the largest user base. A framework for sustainable design and development that is taught in high schools, colleges, and universities could remedy that.

Workarounds
At this point, the best workaround begins with a simple conversation. When agencies pitch new work, why not include a slide on how you are committed to providing low-impact, high-efficiency digital solutions powered by renewable energy? Some clients won't care in the least. Most will want to know if they're more expensive. But some will share your vision and be willing to embrace more sustainable solutions. Find more of *those* clients.

If you don't want to throw sustainability into your pitch mix, why not include it as part of client discussions? Set a page weight budget—where you limit the weight of your pages in kilobytes to a certain size—and discuss the idea during project kickoff and throughout the project. Help them understand that image carousels on the home page don't perform well across devices, convert poorly, and hog bandwidth. Suggest a renewable energy–powered hosting provider.

Similarly, when your software startup begins entering user stories into Trello during discovery sessions, why not add a conversation about green hosting into the mix? Many of the Internet's most successful startups and established media outlets host their products and services with Amazon Web Services (AWS), even though AWS has been slow to adopt renewable energy policies and their historic track record of transparency around these issues has been questionable (though they are getting better). Is there a better, more sustainable alternative for your application? How about Google Cloud?

Finally, if you can't commit to using a hosting provider that is powered by 100% renewable energy, find one that purchases renewable energy credits. Although it's not a perfect solution, as we'll discuss in the next chapter, at least it's something.

Conclusion

In this chapter, we learned more about why Internet sustainability standards are important and why standards themselves could potentially move too slowly to have real impact. We talked about the framework proposed by this book for building more sustainable digital products and services and we covered barriers and workarounds for reaching environmental goals. Using the framework set forth in this chapter and covered in detail in the subsequent chapters, product development teams have the potential to make a significant difference in the energy used by the Internet and its billions of users.

Action Items

If you want to explore the concepts outlined in this chapter before diving deeper in the upcoming chapters, try the following:

- Read James Christie's article "Sustainable Web Design" on A List Apart or Pete Markiewicz's article "Save the Planet Through Sustainable Web Design" on Creative Bloq
- Read the Future-Friendly Web approach to building online websites and applications (*http://futurefriendlyweb.com*)
- Explore some blog posts (*http://www.sustainablewebdesign.org*)
- Watch video presentations on the Sustainable UX YouTube channel (*http://bit.ly/sustainable-ux-conf-2016*)

[3]

Sustainable Components

What You Will Learn in This Chapter
In this chapter, we will cover:

- Why green hosting powered by 100% renewable energy is so critical for building a more sustainable Internet.
- How other components, such as sustainable business practices, Lean/Agile workflows, boilerplates, and software frameworks can make your web solutions more sustainable.
- What the potential barriers are to hosting a digital product or service powered by renewable energy.

A Greener Apple

In the 2012 edition of *Clicking Clean*, its annual report on the Internet's environmental impact, Greenpeace criticized Apple for its data centers' reliance on fossil fuels. Within two years, Apple had publicly committed to a goal of powering its data centers with 100% renewable energy and the company has made great strides in turning that goal into reality. They have published facility-level details on energy consumption and improved the amount of information shared publicly on electricity used at each of its data centers. Apple is also one of the most aggressive among larger Internet companies in pursuing renewable energy to power its data centers, deploying a significant number of solar installations and micro-hydro projects since making the commitment to 100% renewables in 2012. The company has been similarly aggressive in reducing energy and carbon footprint for all of its operations. A data center location in Viborg, Denmark will be designed to capture excess heat and route it into the town heating system to heat other buildings.

Apple is but one example of companies being transparent about their energy use and ambitious in their move to renewables. Others, like Salesforce, Rackspace, Facebook, Etsy, and Google have made similar commitments. This is very encouraging and will hopefully inspire other companies to follow suit.

There is also still a long way to go. For example, many colocation providers—those that provide hosting to millions of websites—still lag far behind their consumer-facing counterparts in commitment to renewables. Plus, pricing strategies and marketing campaigns often drive demand for digital products and services that are powered by fossil fuels. Amazon Web Services (AWS), for example, has priced its cloud infrastructure inexpensively while remaining feature-rich, making it a very attractive option for cash-strapped startups. According to the Kauffman Index, an annual report on global entrepreneurial activity from the Kauffman Foundation, there was an average of 530,000 new business owners per month in 2015.[1] Many of these new businesses, especially those whose products and services are digital, would require easily scalable digital infrastructure and might find AWS to be a viable solution. As we established in Chapter 1, AWS has a less-than-stellar reputation when it comes to energy transparency and commitment to renewables.

The head of a mission-driven startup once said to me, "Show me an alternative that allows you to scale up web servers, databases, or computational power on demand for a similar price point. As it stands, we can pay a few hundred to a few grand per month and be up in minutes, or we could spend an engineer's time building and maintaining our own servers. With limited resources, we don't really have options."

There is, however, at least one comparably priced viable alternative to AWS that has committed to renewable energy (see the sidebar "Google Versus Amazon" on page 69).

---

[1] Ewing Marion Kauffman Foundation, "The Kauffman Index 2015: Startup Activity | National Trends". (*http://www.kauffman.org/~/media/kauffman_org/research%20reports%20and%20covers/2015/05/kauffman_index_startup_activity_national_trends_2015.pdf*)

Google Versus Amazon

Google Cloud Platform (GCP) is a viable alternative to AWS for hosting scalable digital products and services. Here are some quick, high-level comparisons:

- Amazon offers more available services.[1]
- Some of GCP's features—like integrated networking, persistent disk, load balancing, and so on—receive higher marks from reviewers than comparable AWS services.
- Google offers simple pricing with automatic reductions.[2] It has publicly committed to passing along to customers any future price reductions that it achieves through technology-driven advancements.
- Pricing for AWS, on the other hand, can be complicated with multiple complex pricing options. In one comparison, for similar services, AWS was as much as 49% higher than GCP.

Most important for this conversation, as part of Google, which has made public commitments to and is one of the world's largest investors in renewable energy, and also offsets all its electricity use, GCP is carbon neutral, making it the clear winner in environmentally friendly cloud Platform-as-a-Service (PaaS) providers.

---

1   Andrea Colangelo, "Google Cloud vs AWS: A Comparison", CloudAcademy Blog, October 30, 2014. (http://cloudacademy.com/blog/google-cloud-vs-aws-a-comparison).

2   Aviv Kaufmann and Kerry Dolan, "ESG Lab White Paper - Price Comparison: Google Cloud Platform vs. Amazon Web Services", Enterprise Strategy Group, June 2015. (https://cloud.google.com/files/esg-whitepaper.pdf)

FIGURE 3-1.
In 2015, Apple announced $2 billion investment plans for data centers powered by 100% renewable energy in Ireland and Denmark as well as an $850 million investment in a California solar farm

THE CHALLENGES OF BEING A TRULY GREEN WEB HOST

Although all the other techniques outlined in this book can help increase efficiency and reduce energy use, none are as effective at reducing environmental impact as where your power comes from. But not all renewable energy is created equal. Green hosting comes in several flavors, some of which are preferable to others. Whereas some hosting providers have the resources to build their own wind farms or solar arrays, others do not. They can, however, bring renewable energy to their customers through other methods, but these options are not without complications. Add to this the fact that marketing departments often make false claims of 100% renewable energy, further confusing this issue.

Plus, the political climate of many areas does not often favor renewable energy as a superior alternative to fossil fuels. For many data center operators, the local political environment and monopolistic utilities make moving to renewable energy a difficult proposition. Sometimes, the only way companies in certain markets can commit to renewables is to purchase unbundled renewable energy credits—a subpar solution as we shall see later in this chapter.

In the United States, for example, Edison Electric Institute (EEI), a company that represents the majority of US-based, investor-owned utilities, has been urging lawmakers and utility commissions to pursue policies that would increase the cost of distributed solar energy. According to Greenpeace's 2015 *Clicking Clean* report:

> Although EEI claims its attacks on solar power are motivated by concerns for ratepayers, EEI board documents show that utility executives were actually concerned about lost earnings, as the growth of distributed solar shrinks their customer base and obviates the need to build additional centralized generation capacity, which serve as a major source of their profits.

At least 10 states—Alabama, Florida, Georgia, Indiana, Michigan, Oklahoma, Tennessee, Texas, Virginia, and Wisconsin—are actively blocking rooftop solar development through destructive policies, according to a 2016 report by the Center for Biological Diversity.[2] These states account for 35% of total rooftop-solar potential in the contiguous United States, but represent only 6% of installed capacity.

These tactics make it challenging for businesses—including data centers—that want to transfer their operations to renewable energy.

Here's a great example: Canvas Host, a green hosting provider and certified B Corp based in Portland, Oregon, initially signed up for wind energy with Portland General Electric (PGE) in early 2007. "It is our belief it doesn't matter the precise utility providing the power," Canvas Host owner David Anderson says. "They are all faucets pouring electricity into a giant bathtub that is the Pacific Northwest's energy demand."

But Canvas Host ran into a hitch: PGE, it was determined, did not actually provide electricity to its Portland data center. Instead, it was through Pacific Power, a California utility. According to David, Portland is like a giant checkerboard with miscellaneous utilities providing power in

---

2 Greer Ryan, "Report: Sunny States' Policies Block Rooftop Solar", Center for Biological Diversity, April 2016. (*http://www.biologicaldiversity.org/programs/population_and_sustainability/energy/throwing_shade.html*)

a haphazard way to city blocks. As Canvas Host was a building tenant and not a direct customer of either power utility, neither company was willing to directly sell it renewable energy.

"For the time being, we were told, one utility was in breach of federal law, as they are not permitted to sell energy to noncustomers (tenants)," David says. "So by late 2011, we needed to find a new solution to purchase renewable energy."

Thankfully, Bonneville Environmental Foundation had just started up operations, and Canvas Host was able to directly transfer its energy purchase to Bonneville. What's better, Bonneville manages multiple energy products, so David could cherry-pick the exact renewable energy he wanted for his company.

Canvas Host's story is, unfortunately, one that's all too common for not just hosting providers, but any company looking to power its business operations with renewable energy. Stories like these make it challenging for conscientious creators of digital products and services to make educated decisions about their energy choices. Designers and developers must have simple, clear options for hosting the digital products and services they create with clean energy. That is not currently the case, but things are slowly improving thanks to the help of conscientious consumers and companies and the hard work of nonprofits like Greenpeace and Green America.

Todd Larsen and his team at Green America work to create an economic system that supports social justice and environmental sustainability. Helping businesses and consumers understand that a digital footprint is part of an overall environmental impact is a key part of what they do:

> We encourage all businesses and consumers to be aware of their digital footprints. Most Americans are unaware of the enormous energy that is required to power the cloud, store their data, and allow them to stream movies. We also encourage large corporations to take their digital footprints into account as part of their overall climate impacts. And, we are working with consumers to encourage Amazon, one of the largest hosting services in the world, to take steps to increasingly power its servers through clean energy—with a goal of 100% clean energy by 2020. We are focusing on Amazon because, unlike several competitors in the tech world, Amazon was using almost no clean energy a few

years ago. Thanks to pressure from allies like Greenpeace, social investors, and Green America members, Amazon is making progress on the issue. The company has a stated goal of 100% clean energy for its servers (but no timeline as of yet), and has initiated four clean energy projects to power its servers. The majority of its energy is still coming from coal. So we are continuing the campaign.

So how do environmentally conscientious design and development teams make educated decisions regarding how they power their applications? Let's try to make sense of the sometimes complicated options.

The aged grid

To understand where the Web's power comes from and how it gets to our servers, let's first talk about some challenges that existing electrical grids pose. Many of them are more than 100 years old, created at a time when electricity needs were simple. They also don't differentiate between power generated by a wind farm, a hydroelectric dam, a solar array, a nuclear plant, oil, or a coal-powered electrical plant. Electricity from all sources goes into the grid and mixes together, making it virtually impossible to know whether the energy that ends up at your home or business came from clean, renewable sources or fossil fuels.

This poses challenges for consumers or companies that want to commit to renewable energy. Currently there are only a handful of options. Here are some of the ways in which companies use market-based solutions to procure renewable energy:[3]

On-site investment

A solar array on the rooftop or a wind farm in a nearby field, these investments are the most straightforward to assess in terms of impact. Most data centers require significant amounts of energy, however, meaning that these investments might only provide a small percentage of a provider's electricity needs. For many smaller providers, this option can also be cost-prohibitive.

---

3 Greenpeace, *Clicking Clean: A Guide to Building the Green Internet*, May 2015, p. 29.

Power purchase agreements (PPAs)
>Hosting providers can secure long-term contracts (often 10 to 20 years) with energy companies to provide renewable power for their data centers. The benefits of this arrangement are twofold: energy companies can use this purchase guarantee of both energy and renewable energy credits (also known as bundled RECs) to secure financing to drive development of additional renewable energy infrastructure while data centers can negotiate a price guarantee on electricity, thus protecting against future rate increases.

Renewable energy credits (RECs)
>Also called guarantees of origin (GOOs) in Europe, RECs are created when renewable energy is generated. They are used to grant the environmental benefits of said energy to the purchaser. These credits can then be bought and sold, but they are not bundled with the actual electricity that was generated upon their creation, so it can be a bit of a shell game due to flooded markets driving their prices down. Because of this, unbundled credits don't typically displace electricity generated by fossil fuels or drive demand for more renewable energy, making them a less viable option for moving us collectively toward a renewable energy–powered Internet.

Direct access
>Some deregulated markets give consumers the choice of electricity provider and provide options for power generated from renewable sources. This direct access, however, is not available in many markets and can be restricted to a limited number of customers or providers.

Green energy tariffs
>Larger data centers can sometimes purchase 100% renewable energy products directly from utilities rather than through a third-party provider via programs known as green tariffs. This is rare, however, and in some of the few markets where green tariff programs have been piloted they have proven ineffectual, with customers incurring significant premium costs and "administrative fees." For these reasons, green tariffs are not viable for most renewable energy–focused hosting providers.

The preceding five examples overly simplify a complicated landscape for powering your digital products and services with renewable energy. The reality is much more complex. Policy changes are still required in many markets to make renewable energy a reality. Also, hosting providers, data centers, and other companies will not aggressively pursue options for renewable energy in their market unless they know there is consumer demand for it.

To complicate things even further, marketers often rely on the complexity of this situation to claim 100% renewable energy without putting the legwork into actually achieving it. Few consumers will challenge such a claim. Making said claim is not as simple as buying unbundled RECs and stating that you are powered by 100% renewable energy, though in many markets there are few other options. One hosting provider even claims in its marketing materials that it plants a tree for every new account, which is wonderful, but doesn't move us closer to an Internet powered by renewable energy. All this further confuses an already overwhelmed consumer base. It is no wonder that clean energy is rarely part of the decision-making process when web teams review where to host their applications.

RECs versus renewables

Given the complicated landscape of powering servers with renewable energy, what's an acceptable solution?

Shawn Mills of B Corp Green House Data says, "It's largely unreasonable to expect a data center to purchase direct renewable energy, so RECs are a good way to encourage growth in the sector. Through them, we are able to certify that the amount of energy consumed by our data centers has an equivalent amount of renewable energy added to the grid."

David Pomerantz of Greenpeace doesn't necessarily agree, because RECs can be unbundled from the electricity that produced them, which can be problematic. He notes that because the market has driven RECs prices so low they often end up not contributing to production of more clean energy, which is what really helps directly address the problem of climate change. "If you're paying for the electricity along with the RECs," he says, "it's a viable solution. But if you're just buying the RECs alone, you're not actually improving the situation. Companies that purchase RECs and not the electricity that comes with them are, in effect, just buying PR."

"In some ways, [RECs] have made the rapid growth of clean energy possible, and used correctly, they can be advantageous," said Green America's Todd Larsen. "At the same time, RECs can be problematic and actually work against creating local, truly green sources of energy."

One of the biggest challenges with RECs, Todd notes, is that all forms of green energy, including the direct purchase of renewable energy from a local provider and RECs—which can be green energy certificates that support a mix of truly clean energy and less green technologies, and can be purchased for energy that would have been produced absent the REC—are often valued equally, when they should not be.

Rob Stevens of ClimateCare, a B Corp based in the United Kingdom that has improved the lives of more than 6 million people and reduced 16.5 million tons of CO_2, notes that there are similar challenges in Europe. "Guarantees of origin, or GOOs, are related to laws which oblige energy providers to get 20% of their energy production from renewable sources," he says. "The market is based on legislation and compliance. Corporations can buy GOOs in Europe to demonstrate that purchased electricity comes from renewable sources. Interest in this is growing, but complicated actions can stifle corporate involvement. It is important to keep communication around this simple and engaging."

David's advice to people vetting hosting providers is that "as a general rule, if they're not talking about their energy supply, they are likely just using RECs."

David, Todd, Rob, and Shawn all agree that transparency and putting the extra effort to not take the easy road (in this case, unbundled RECs) will go a long way in *all* sustainability efforts, but especially in sourcing renewable power.

"We specifically purchase wind energy RECs that go toward production of local projects," Shawn says. "RECs can be used to directly fund wind turbine construction, for example. Energy companies can apply to receive funds from RECs and use them for the design, construction, or ongoing operation of a renewable energy production facility."

Green House Data chose fellow B Corp Renewable Choice because it is a well-established broker with a history of successful investment in renewable energy across the United States, where Green House Data is located. It added some RECs from WyoREC to support REC development in its hometown of Cheyenne, Wyoming. Renewable Choice does projects in the region, but this was a new local venture. "EPA recognition is also vital for us," Shawn says, "as we are an EPA Green Power Partner. The RECs have to be tracked in order for us to feel comfortable that their funds are going towards legitimate projects."

For many companies, RECs might be the only affordable option. "In that case," David says, "buy the highest quality RECs you can. Get engaged with the local utility. Push your suppliers. Take the high road. This is an area where smaller companies can hopefully get resources through organizations like B Lab and Green America."

For companies that own and operate their own data centers, Greenpeace expects them to advocate for greater renewable energy supply, and for greater access to renewable energy. That could be with federal policymakers, state policymakers, state regulators (like public utility commissions), and directly with the utilities themselves.

"Companies who do not own their own data centers don't have as direct a relationship with the utility," David says, "which makes that a bit more challenging. They can and absolutely should engage in political advocacy for renewable energy if they are committed to moving their operations in that direction. But we encourage them to leverage relationships where they'll have the highest impact, and that's often with their vendors. So if their company's data is hosted by a third-party cloud provider, or in a colocation data center, we encourage them to advocate with those parties to offer them renewable energy options, to improve their transparency, to do their own advocacy up the line, and so on."

David offers other advice, as well: customers of any of these companies can and should communicate their desire for a green, renewably powered Internet. They can also do their best to power their own homes and footprints with renewable energy, and like everyone else involved, register their voice with policymakers so that it becomes easier for them to go green, too.

Companies that create digital products and services are in a unique position to advocate for green hosting that is powered by renewable energy while also offering the same simplicity, uptime, commitment to customer service, and quality that their fossil fuel–powered counterparts do. Of course, this goes hand in hand with what your resources allow. When I moved my personal blog from a provider that had hosted it for nearly five years, I told the company owner that I needed to house it with a provider that uses renewable energy. He said he didn't know that was even an option and mentioned he would look into it. I sent him some links. Sometimes, it's as simple as having a conversation, something all designers and developers could easily have with their providers.

"Widespread renewable energy will be the single biggest game changer for this sector," says Shawn. "If we can get everything on renewables, any energy wasted from inefficiencies has less of an impact.

"From there we can work on getting all the tiny data centers scattered across the globe into better operational shape. Those are the real drains on resources—there are little 5,000-square-foot facilities that probably have a comparable carbon footprint to one of Google's 100,000-square-foot buildings, just because of outdated design, equipment, and management."

There is room for other companies, including smaller hosting providers like AISO.NET and others, to be leaders in this area alongside the big players that might have more resources. Their customers—like designers, developers, and web teams—can help them by making it known that there is interest in reliable, efficient hosting powered by 100% renewable energy.

To grid or not to grid?

Data centers or hosting providers that currently wish to generate their own renewable energy will find themselves taking on a serious commitment of time and money, something that might be out of reach for many smaller companies that want to provide green hosting but don't necessarily have the resources to do so. Though pricing for solar is coming down, powering your data centers using renewable energy requires investments in hardware that needs to be maintained and/or upgraded. In many markets, it also means trudging through regulatory red tape and crossing legislative hurdles before you can even begin the planning process. Putting a solar array on your rooftop or a windmill in the parking lot, however, also guarantees that you are getting energy from a local, renewable source. Because the source of your power is nearby, it can also mitigate transmission loss that comes with using the grid.

Renewable energy can be installed with or without access to the larger electrical grid:

A *microgrid* does not rely on grid power at all. This small-scale power grid can operate independently or in conjunction with the area's main electrical grid. Any small-scale localized station with its own power resources, generation and loads, and definable boundaries qualifies as a microgrid. This is not often realistic for data centers, which require large amounts of electricity. Plus, because reliable access to data is critical to their business model, a data center that relies on a microgrid would also require a reliable backup power supply, which is likely attached to the grid.

Data center wind farms or solar arrays can also contribute power back to the primary electrical grid, which can then be used by customers of that grid. Regulations on how you are compensated for the energy you supply the grid vary from market to market.

Winnipeg-based Canadian B Corp and web design firm Manoverboard noted in its Green Web Hosting whitepaper,[4] "Even for most data centers with on-site or off-site renewable energy, the center maintains a tie to the local electricity grid and the power its equipment uses is from these grid-fueled power plants. This local electricity mix can rely heavily on fossil fuels that produce greenhouse gases (GHGs)."

Good company versus good marketing

Why does this matter? Because where you host your digital product or service makes a difference. It can be challenging to find quality sustainable solutions, but those seemingly small decisions can add up.

---

4 Andrew Boardman, "Green Web Hosting and Environmental Impact", Manoverboard, December 5, 2015. (*https://manoverboard.com/green-web-hosting-and-environmental-impact*)

There is a common question asked by B Lab, the organization that certifies B Corps: "How do you tell the difference between a good company and just good marketing?"

For green hosting providers, the answer lies in transparency about where their energy comes from and how they get it. "Whatever angle on sustainability a company chooses, it should be transparent and honest in what it is doing, who it is benefiting, and it should of course mean something to the consumer," says Canvas Host's David Anderson.

Redefining Success in Business

FIGURE 3-2.
What is the difference between a good company and just good marketing?

Canvas Host developed "meaningful metrics" to use in its own program. "For us, we wanted to not only offset our energy consumption, we wanted to measure it to demonstrate ongoing progress," David says. The company established these three metrics for its sustainability program:

Total amount of electricity consumed

As a company grows, it will use more energy. Through sustainability efforts, Canvas Host's energy consumption plateaued in about 2010–2011. Since then, it has steadily declined. It is now at about 14,000 kWh per month, far below its one-time peak of 22,000 kWh of consumption.

Average power consumed per service line

There is a huge range of energy consumed depending on whether a customer is on a shared server (using 1/200th to 1/400th of the server's resources), versus a dedicated server (using 100% of the server's resources). Canvas Host wanted to track shared, dedicated, and virtual private servers. By 2008, it had rolled out a line of mini green servers, which have again turned the metrics on their head. Its latest servers, based on the Intel NUC, draw perhaps 1/10th of 1 Amp of power; yet they have performance rivaling huge dedicated servers. Combining one and two, you end up with Canvas Host's most important metric.

Number of domains hosted per Amp of electricity consumed

As the logic follows, if domain density increases, it means the operations are more efficient without increasing power consumption. When Canvas Host first began, it could host about 30 to 40 domains per Amp. They're now close to 600 domains per Amp.

"It's simply incredible how far this has come," David says. "And yet, the story is still evolving."

Sustainability initiatives also *must* be balanced with reliable service. All the wind power in the world won't make a difference if a company's servers consistently go down or they haven't committed to a stellar customer service experience. People will flock away from them in droves.

Andrew Boardman from Manoverboard, had this exact thing happen when one of his clients who cared very much about sustainability chose not to go with a green provider. "We reviewed the various options available, including using a relatively new green hosting vendor, staying put, or moving the site to a highly reputable data provider with no sustainability policy," Andrew says. "When we got to the other side of the tracks, together we chose the latter—performance and support won over renewables and e-waste. I think we were all a bit disappointed, but no one was prepared to risk a very valuable project with a smaller, newer data provider—green or not. The client can use renewable energy credits to offset their choice of a nongreen provider."

The best web hosts offer the complete package: great, feature-rich hosting at competitive prices, excellent customer service, *and* servers that run on renewable energy. Unicorns, in other words.

We will cover green hosting barriers, workarounds, *and* unicorns later in the chapter. Green hosting is only one sustainable component in a list of many. Let's explore some of the others.

Other Sustainable Components

Even though hosting is the most important component to directly contribute to a digital product or service's environmental footprint, it's a far cry from being the only one. There are others, as well. From workspaces and workflows to frameworks and mission statements, these components bring context, purpose, and efficiency to the process of creating more sustainable digital solutions. In the following sections, we'll review a few of these.

ENVIRONMENTALLY FRIENDLY WORKSPACES

Environmentally friendly offices help you to get things done more efficiently, they use less energy, and they produce less waste. No surprise there. From better air and light quality to more efficient power systems and better control over one's own personal space, an environmentally conscious office also encourages teams to make more conscious choices in what they do every day. This is where sustainability practice from architecture, interior design, and construction can be directly applied to web design and development.

FIGURE 3-3.
URL the earthworm—compost produced by Mightybytes worms is used to fertilize plants around the office, including the two living green walls

Mightybytes has taken numerous steps to create a work environment with less impact that helps the team make more sustainable choices every day.

- There are recycling bins near every desk and in the company kitchen.
- Food waste is composted using onsite vermiculture (worm composting). The resulting compost fertilizes office plants and, at times, employee gardens.
- Living plant walls produce oxygen, improving air quality.
- The HVAC system is scheduled to be on when there are people present and off when there aren't. It also tracks energy use from month to month, giving us goals to benchmark over time.
- Small, desk-side space heaters give individuals greater control over their personal space while reducing overall energy used by the company's heating and cooling system.
- All lights are LED rather than more energy-intensive incandescent or even mercury-filled CFL bulbs.
- The walls are made from partially recycled materials.
- Telecommuting is encouraged, as are more sustainable modes of transportation like cycling or public transportation.

These choices put sustainability front and center in the ongoing dialogue of how we run our company. Working in a more sustainable office isn't the least bit revolutionary. Companies have been doing this for decades. It's important to note, however, that folks who work in more environmentally conscious offices tend to make more environmentally conscious choices in their personal lives, as well, sort of a reverse Jevons paradox, if you will.

"Sustainability plays a big role in our company," says B Corp LimeRed's owner Emily Lonigro-Boylan. "We're interested in continuous deployment—in making our business better and building projects. It's a lot less effort to be constantly improving a process than to reimagine it once a year."

In terms of some of the physically sustainable areas LimeRed has control over, the company recycles, it keeps the heat low, it orders all of its office supplies carefully from another B Corporation. "It's tough when you rent, but we do what we can."

Emily also notes that LimeRed is conscious of the amount of data they're creating. They try to execute projects in the most efficient, minimal way. "If there's a way to simplify, do it. If there's a way to share, do it. If there's a way to build on what someone else has already created, do that, too. There's no sense reinventing the wheel if something already exists that does a good job."

Emily also notes that one big thing all companies can change right now is being more aware of the digital footprint of files and keeping them clean. "We did a site audit before a big project last year and we found pieces of the site on four different servers, old hidden pages and loads of files in storage just taking up space," Emily says. "More people who maintain sites aren't developers, and this kind of waste needs to be stopped. Imagine how much old stuff exists just sitting on servers that can be deleted."

Environmentally friendly workspaces and clean file servers are only part of the picture. Some companies extend sustainable philosophies to the very core of their DNA.

THE STAKEHOLDER MODEL

In the Preface, we talked about how being a certified B Corp drove Mightybytes to reevaluate its own supply chain and led to many process efficiency improvements, not to mention the creation of our web sustainability tool, Ecograder, and the general philosophy for how we approach business.

This rapidly growing global community of businesses has adopted the stakeholder model philosophy, where the needs of all stakeholders, not just shareholders, are carefully considered when making business decisions. In this model, a stakeholder is defined as any group or individual who can affect or is affected by the achievement of the organization's objectives.

Though the idea isn't new—stakeholder theory can be traced back as far as 1963 at the Stanford Research Institute (SRI)[5]—this approach to doing business has in recent years gained significant amounts of traction. Also commonly called *triple-bottom-line businesses*—those that focus on people and planet alongside profit—B Corps are only a part of a larger global community of companies concerned about the role business plays in the world. Co-ops, fair trade, and conscious capitalism are a few other examples.

This is significant because as Chapter 1 points out, innovation and disruption often occur in companies that consider stakeholder needs and the environment alongside their need to make profit. A company that consciously considers such things will be more likely to examine the environmental impact of its digital products and services than one that doesn't. From this philosophy, groundbreaking innovations can arise.

SUSTAINABLE MISSION STATEMENTS

A company's mission statement is the rallying cry around which its entire existence grows and thrives. In addition to defining the company's purpose, a mission statement helps all company stakeholders get on board with its vision. When sustainability is included in these statements, it helps weave the concept of making more sustainable choices into everything the company does.

When devising a company mission statement that is driven by sustainability, here are some questions to consider:

- Is the mission statement easy to understand so that stakeholders can rally around it?

- How can the company make sustainability a key brand attribute?

- How can the company help its teams make more sustainable choices in their work?

- How will the company educate clients and other stakeholders about the new mission?

---

5 Philip Webb, "The Origins of the 'Stakeholder' Concept", TAM UK – Organisational Strategic Planning Specialists, September 20, 2013. (https://tamplc.wordpress.com/2013/09/20/the-origins-of-the-stakeholder-concept)

Here are some examples of company mission statements from several industries that include sustainability or a commitment to the environment:

Mightybytes, interactive agency based in Chicago
"Through a range of product and service offerings, provide creative, technical, and marketing expertise to help conscious companies solve problems and achieve measurable success. Our company vision and friendly, collaborative environment are guided by a deep commitment to the triple bottom line of people, planet, and prosperity."

Dolphin Blue, online retailer of environmentally friendly products
"To serve as your trusted source of the world's most environmentally and socially responsible products—to assure your well-being and a healthy, sustainable planet."

Gelfand Partners, San Francisco-based architecture firm
"At Gelfand Partners Architects, we seek to increase the impact of architecture on the most important issues affecting life in the designed environment. Our mission is to design sustainable buildings that serve our diverse and dedicated clients, encourage healthy personal and community life and develop positive impact on climates and habitats."

W.S. Badger Company, natural products brand
"To create fabulously pure and effective products of the highest natural quality, based on simplicity and thoughtful preparation, with the intention to protect, soothe and heal. To run a business that is fun, fair, and profitable; where money is fuel, not a goal; and where our vision for a healthier world finds expression through the way we work and through the way we treat each other and the people we serve."

Cabot Creamery Cooperative, a Vermont-based dairy co-op
"Our credo is 'Living within our means and ensuring the means to live.'"

Jed Davis, director of sustainability at Cabot Creamery Cooperative says that while its credo might initially seem simple, a lot of thought went into it. "This ties out very intentionally to our vision and approach for sustainable growth," he says, for the following reasons:

- It is based squarely in capital theory and an understanding of the importance of managing vital capitals.
- It is also very much aligned with the principles of cooperation, as outlined by the International Co-operative Alliance (ICA).
- It is also based on an approach that is stakeholder-centric.
- Finally, it is based squarely on the notion of well-being: the vital capitals, whether natural or anthropogenic, need to be managed by us for their impact on the well-being of our stakeholders.

Jed notes that these are an evolution of the original Rochdale Principles of Cooperation dating back to 1844.

LEAN/AGILE WORKFLOWS

Traditional "waterfall" workflows modeled on 20th-century assembly lines are bad for sustainability. Streamlining workflows can lead to fewer resources used during a project's life cycle, which can in turn serve as a potentially more sustainable approach to said project. With a focus toward ongoing collaboration and communication, creating highest-value deliverables first, and minimizing waste, Lean or Agile workflows can significantly influence the resources needed to execute a project.

"In many cases, the outcomes people want can be delivered in a number of ways beyond what is specified by a client's set of requirements, and these can often be simpler to deliver," says Chris Adams, founder of Product Science, a UK-based agency working primarily with organizations that are addressing social or environmental problems as part of their business. "During a project, it's common to learn new things about the organization or the problem, which make some parts unnecessary—by working in an incremental, iterative fashion, you retain the ability to act upon these, reducing over the length of the project."

Going over the waterfall

Many projects focus on extensive (and often unnecessary) front-loaded scoping and specification documentation, much of which is tossed as new project requirements are learned throughout each phase of execution. This phase-based, requirements-heavy approach, which frequently requires that one phase be completed before beginning another is typically referred to as a *waterfall method*. Problem is, this approach doesn't always account for new ideas that come up as the project progresses. Changes in scope or definition of a project can have profound effects on how the final deliverables are received by end users. Wasted efforts or miscommunication that can happen in the translation between extensive specifications created up front and actual tasks during project execution can result in mismanaged resources, which affect not only the company's financial bottom line, but its impact on people and the planet, as well. Let's explore what this means.

With a waterfall-based approach, teams with different specialties work in silos, rarely collaborating with other teams. Sets of deliverables are lobbed over a (figurative) wall to be completed by the next team. Waterfall projects can promote waste in several ways:

- The process of collecting many specifications up front doesn't account for things that might be learned during project execution. Oftentimes, specifications need to be thrown out or significantly revised as the project progresses.

- Their rigid structure doesn't typically allow for shifts in scope, even when those shifts could potentially affect project outcomes like budget or timing on deliverables.

- Executing a project in a linear rather than iterative manner means project teams work in silos, resulting in less cross-team collaboration and communication.

The cone of uncertainty

Related to waterfall is what is referred to as the *cone of uncertainty*, which purports that unknowns at the beginning of any project require that estimates to execute the project be as much as four times higher than one might expect. As a project progresses, it has a diminishing range of uncertainty. Because budgets and timelines in waterfall projects are based on specific deliverables, clients write exhaustive requests for proposals (RFPs) that outline as many project variables as possible.

Web firms (or any service provider working on large projects, for that matter) negotiate in a time- and resource-intensive process that generally yields inaccurate results because the bidding agency often quotes high to account for the cone of uncertainty. The cone of uncertainty concept originated pre-Internet, in 1950s engineering and construction management, but it definitely applies to today's web projects. Of course, not every waterfall project succumbs to these potential pitfalls, but many do.

Alternatively, what if there were a way to get projects done faster and potentially cheaper? Wouldn't that tend to be a more sustainable choice?

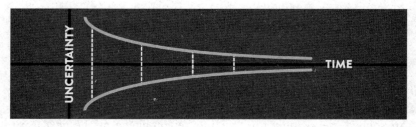

FIGURE 3-4.
Estimates at the beginning of a project are considerably more than at the end because as a project moves toward the deadline, specific deliverables become known

Agile methods

A 21st-century solution to the problems of assembly-line waterfall methods, Agile promotes collaboration over documentation. It values continued learning and embraces conflict. It focuses on delivering the highest-value features first. All of this can result in resources being managed more effectively.

There are many variations of Agile across industries (Scrum, Lean, extreme programming, pair development, Kanban, etc.) each with its own unique variation on what is essentially an interactive and collaborative process for producing projects (or products) faster with fewer resources. Though similar systems were developed for manufacturing as far back as 1948,[6] it wasn't until the publication of *The Agile Manifesto*

---

6 Wikipedia, "Toyota Production System". (*https://en.wikipedia.org/wiki/Toyota_Production_System*)

in 2001 that Agile methods came into their own for the software development industry. Today, agility concepts are applicable across a wide array of industries, markets, and processes. In the world of digital products and services, Agile methods are seeing a steady rise in adoption rates across startups and digital agencies.

"One of the things I love most about Agile is that we're constantly learning and asking questions about it," says Lime Red's Emily Lonigro-Boylan. "Once a client said in a retro (the meeting at the end of a two-week building period called a Sprint): "You mean I talk about what went wrong now? I don't wait until the end of the project?" My team said "Yeah! Of course we want to be able to fix any issues now, not wait!" Being agile means you're aware of the minutiae of the project and the overall goal at the same time." Emily notes that her clients and team working seamlessly is what makes their projects work. "Whenever they need to make a quick decision, they've built the rapport and trust to do it. It's amazing to watch. I can't believe we ever did it any other way."

There are a few things that make Agile methodologies potentially more sustainable than their waterfall counterparts:

- Agile methods require ongoing collaboration from key team members, including those on the client side. More consistent communication yields better results faster.

- Agile focuses on producing deliverables of highest value first, leaving those with less value to later in the project life cycle (or not at all if time and budget constraints take precedence). This produces less waste.

- By focusing on continued learning rather than trying to define everything up front, Agile teams can quickly pivot from features that aren't working, reducing time spent on those features.

- Agile web teams incorporate user testing directly into their process rather than outsourcing it or saving it until the project's end. This identifies bugs or problematic features earlier, when it takes less time to fix them.

- Scaling in both directions—up and down—is built into agile methods.

The Lean Startup movement is an example of agile methods reshaping how businesses bring products to market, letting them focus only on the minimum number of features a product needs to be considered functional, also known as a minimum viable product (MVP). By not adhering to the "more is more" philosophy of application design and development, practitioners of agile methods, including those in the Lean Startup community, create more useful products and services with less waste.

Agile resources

While a deep dive into Agile methods is beyond the scope of this book, the following resources cover both Agile software methods and organizational agility:

- Andrew Stallman and Jennifer Greene, *Learning Agile: Understanding Scrum, XP, Lean, and Kanban* (O'Reilly, 2014).
- Pamela Meyer, *The Agility Shift: Creating Agile and Effective Leaders, Teams, and Organizations* (Bibliomotion, 2015).
- Tim Frick, "Go Agile, Build Better Digital Products and Services" Mightybytes Blog, February 24, 2016 (*http://bit.ly/1Wj9bXO*)

SOFTWARE FRAMEWORKS

Software frameworks like CodeIgniter, CakePHP, or Ruby on Rails help developers save time when creating web applications by offering a collection of tools that automate common development tasks. CSS frameworks such as Bootstrap, Blueprint, or Cascade apply this same idea to frontend development. Although they save time, some of these frameworks also contain unnecessary overhead that can slow page performance or download times. Choosing a framework that is more sustainable for your organization is dependent on a few things:

- Your team's skill set and what they are comfortable with (or willing to learn).
- The specifications of your application: how much of the existing framework will be unnecessary overhead for what you're building?
- The efficiency of both the existing code within the framework and that which you put in it yourself.

As an example, Dr. Pete Markiewicz, an instructor in interactive and web design at the Art Institute of California, Los Angeles, has been working on a Green Boilerplate, an HTML5 template that minimizes CPU grind and resource consumption for mobile and desktop computers.

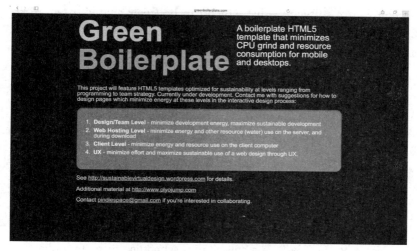

IMAGE 3-5.
A Green Boilerplate can potentially help your site minimize energy use

The Green Boilerplate aims to minimize energy use in several areas:

Design team level
> Minimize development energy, maximize sustainable development.

Web hosting level
> Minimize energy and other resource use on the server and during download.

Client level
> Minimize energy and resource use on the client computer or device.

UX
> Minimize effort and maximize sustainable use of a digital product or service's design through better UX.

Web frameworks like the Green Boilerplate—built to maximize the user experience while minimizing resource use—offer opportunities to help designers and developers build more sustainable digital products and services. We cover frameworks in more detail in Chapter 6.

OPEN SOURCE AND SUSTAINABILITY

Open source software—that which can be freely shared, used, modified, distributed, and improved to promote universal access—has its own ramifications for sustainability. There's the sustainability of the open source movement itself and the efficiency of the open source code the community generates. There is also great potential in sharing resources in open, collaborative environments that can result in much more sustainable solutions than their proprietary counterparts. Think GitHub, Uber, TaskRabbit, and so on.

Mike Gifford, owner of certified B Corp Open Concept Consulting, notes that using open source tools can be considered more sustainable than their proprietary counterparts for several reasons:

> Open source tools are driven by the community of users and developers. When they have a critical mass supporting them, they offer unparalleled opportunity to lean on other developers to develop, maintain, and extend a code base. The transparency of open source software communities helps to ensure that developers are building with best practices in mind.
>
> All code is debt. If you have developed a custom application or are leveraging proprietary code, the number of people who are able to review it is limited. A lot of developers have a desire to write a solution from scratch in the latest cool code base. The more people review a code base, the better it will become.
>
> Many open source tools have very wide adoption. If performance improvements are made in these core libraries, they will be picked up by thousands if not millions of other applications.
>
> One of the concepts adopted by many free software tools is to be modular. Ideally, you can only run the software you actually need to run (and no more). This is great for security, because really why would a web server need to have a GUI, but there are also advantages for sustainability. All that extra code takes resources to run.

A more sustainable community

It's hard to invest in a solution if you don't know whether it will be around in a few years. This has been an ongoing challenge in the software industry for decades. Programming languages—whether proprietary or open source—are replaced, improved upon, or support is discontinued. Finding a developer who knows a proprietary legacy system can be time consuming and costly. Conversely, it is sometimes easier (and potentially more cost effective) to find a designer or developer familiar with open source tools.

Thousands of people maintain the most popular open source projects, for example. Although bugs or security vulnerabilities in open source projects can be exploited by hackers, there are also significant efforts afoot to ensure those things don't happen again.

"People should think about open source like free kittens," says Mike Gifford. "They are amazing, but if you don't take care of them, they can be hell."

The movement itself has run into challenges in being self-sustaining, as well. According to the Technology Innovation Management Review:[7]

> In 2008, many thought the open source movement could not survive the widespread adoption of open source software without commensurate contributions back, whether in code or cash. Since that time, however, open source has flourished, and it has become robustly self-sustaining. This dramatic improvement in the health of the open source ecosystem derives from two primary trends: a move toward more permissive, Apache-style licensing, coupled with an increase in open source contributions from web technology companies like Facebook.

It is encouraging to see larger companies like Facebook embracing open source principles and practices. Coupled with better licensing practices, this can give the open source movement the progress it needs to be more sustainable for the long term.

---

7 Matt Asay, "Q&A. Is Open Source Sustainable?", *Technology Innovation Management Review* 3(1): 46-49. (*http://timreview.ca/article/650*)

Hacking the climate with open source

It is possible for an open source project to become inefficient or fragmented as a result of there being too many contributors. But it's also possible for open source platforms to address challenges like resource utilization, excess capacity, diversity, scaling, innovation, exponential learning, and yes, even saving the planet.

In a 2015 presentation at Linuxcon, Zipcar founder Robin Chase noted, "Industrial capitalism is dead because the Internet exists and sharing is a better way of extracting more value."[8] Bold claims. Zipcar and Airbnb were examples she gave of open source collaboration defying the laws of physics: "We can build the largest hotel chain in the world with Airbnb in just four years." Tesla, creators of popular electric cars, also plans to embrace open source solutions. Although it remains to be seen whether Airbnb will save the planet, Chase does bring up a good point that open source platforms offer opportunities to redistribute wealth and change entire industries without incurring the planet-damaging overhead that would come with, for example, *actually* building the world's largest hotel chain. Chase also highlighted that shared network assets deliver more value than closed assets and just as open collaboration often yields superior results to closed or proprietary solutions due to the sheer number of people collaborating on an idea.

GitHub, for example, has more than 30.7 million code repositories uploaded by its users for all to share and alter.[9] The company, which began as a weekend project, has been evaluated at $750 million and adds about 10,000 new users on average every weekday.[10] The White House, Facebook, Amazon, and LinkedIn all use it for various projects, as do about 12.1 million others. It offers a vast resource for developers to test out new ideas and find resources for their projects.

This is important to consider as we weigh the sustainability pros and cons of the applications we build. More users can mean more environmental impact on the backend and frontend of the applications

---

8 Sean Michael Kerner, "Linuxcon: Open Source Peer Collaboration Could Save the Planet", Datamation, August 17, 2015. (*http://www.datamation.com/open-source/linuxcon-open-source-peer-collaboration-could-save-the-planet.html*)

9 GitHub, "Press". (*https://github.com/about/press*)

10 Vijith Assar, "The Software That Builds Software", *The New Yorker*, August 7, 2013. (*http://www.newyorker.com/tech/elements/the-software-that-builds-software*)

themselves, but conversely, distributed collaboration offers ways to accomplish tasks with fewer dependencies and bottlenecks, making it possible for us to build applications via a process which is in and of itself more sustainable.

WEB STANDARDS

Web standards exist for many reasons, not the least of which are to make the Internet more accessible and readily available to all and to make your work more future-friendly inasmuch as can be expected in an industry that changes every hour.

For web designers, much of this standards work comes in the form of separating structure from style and behavior. As standards leader Jeffrey Zeldman sums up on his blog, "This separation is what makes our content as backward-compatible as it is forward-compatible (or 'future-friendly,' if you prefer).[11] It's the key to reuse. The key to accessibility. The key to the new kinds of CMS systems we're just beginning to dream up. It's what makes our content as accessible to an ancient device as it will be to an unimagined future one."

Because of the inclusive, accessible, and future-friendly nature of web standards, adhering to them when designing and developing web applications and digital products or services is typically the more sustainable choice.

Not sure if your application is standards-compliant? Here are a few easy things you can do to find out:

- Run it through the W3C's Markup Validation Service (*https://validator.w3.org*) or use the Firefox HTML Validator add-on (*https://addons.mozilla.org/en-US/firefox/addon/html-validator*)

- Check how compliant your work is with accessibility standards using Formstack's 508 Checker (*http://www.508checker.com*)

- Still looking for other ways to check how standards-compliant your code is? Check out the W3C's complete list of tools (*http://www.w3.org/WAI/ER/tools*)

---

11 Jeffrey Zeldman, "Of Patterns and Power: Web Standards Then & Now", Zeldman.com, January 5, 2016. (*http://www.zeldman.com/category/web-standards*)

Chapter 6 covers the advantages of web standards in building more sustainable applications in greater detail.

Potential Barriers and Workarounds

Keeping life cycle assessments in mind, the components that go into your digital products and services reflect how sustainable they are overall. Not all digital products or services are accessible, open source, standards-based, hosted on servers powered by renewable energy, and created by using Agile methods by an environmentally conscious company that commits to all its stakeholders, not just its shareholders. In fact, currently very few are. But this is something we can aspire to. Potential barriers lurk around every corner though.

Take green hosting, for example. Although there are some options for hosting your digital products and services with data centers powered by renewable energy, the quality and service offered might not yet match the levels you've grown accustomed to from working with larger companies.

A TALE OF GREEN HOSTING WOE

This story sheds light on the barriers and workarounds that impede the growth of an industry that could lead the way with renewable energy: Mightybytes' search for one of the mythical green hosting unicorns mentioned earlier in this chapter.

In 2010, we set out on what would turn into a years-long quest to find a company that blends commitment to renewable energy with a dedication to customer satisfaction and high-quality, reliable web hosting services for our project needs. As a B Corp whose supply chain's biggest environmental impact is the electricity used to power the solutions we build, hosting with renewable energy is the only logical option.

We thought it would be a short, simple quest, but, boy, were we wrong! From late 2010 to 2015, we started and ended relationships with nearly a dozen different green hosting providers whose energy mix ranged from direct sources of renewable energy to RECs.

Most of these relationships were ended because they missed two key components:

- Reliability
- Good customer service

FIGURE 3-6.
There are many green hosting providers out there, but sadly very few unicorns

Site reliability and customer service (or lack thereof) proved to be the primary factors that caused us to sever relationships with green web hosting providers, again and again. In some cases, there were also performance issues, but overall, uptime consistency and bad customer service proved to be the deal-breakers. Our customers' websites went down repeatedly and sometimes inexplicably. At best, customer service representatives either referred us to the company's knowledge base or recommended expensive add-ons. In some cases, calls simply were ignored. Meanwhile, our clients were understandably freaking out about their websites going offline. Rather than partner with us to provide reliable service and create mutually beneficial relationships based on trust and a commitment to help each other, these providers instead lost business.

What these companies had in their commitment to the environment they seemed to lack in these other areas. Based on several years of auditing green web hosting companies, here are a few suggestions on how they might improve their businesses to better serve customers while simultaneously growing the renewable energy–powered Internet.

Reliability

By far, the most common problem we experienced when working with these companies was reliability in site uptime. Websites went offline unexpectedly and repeatedly from provider to provider.

Thomas Burns, cofounder and vice president of business development at Green House Data notes that reliability is one of his company's highest priorities:

> "Our facilities have multiple levels of power infrastructure redundancy," he says. "We use the term '2N' and 'N+1' to describe our power infrastructure design methodology. The 'N' refers to need. For instance, we have designed our UPS (Uninterrupted Power Supply) system to be a 2N system. That means we have built '2' times the need (N) for resilience. If we lose a battery in our UPS there are two more batteries to replace its function. We do this on all aspects of our power infrastructure deployment from power to the rack through the generators and utility feeds."

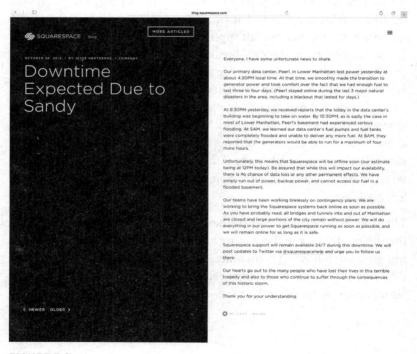

FIGURE 3-7.
Stellar customer service in the face of climate change from Squarespace

Of course power sources aren't the only thing that can bring a website offline, but it is good to see a company committing to reliability at that level.

Squarespace's efforts to keep its 1.4 million websites online during Hurricane Sandy is a great example of one company's commitment to reliable hosting services. During the disaster, the company got permission to carry fuel into their data center in plastic bottles to keep the servers running. Several team members stayed on site during the storm to do so if that was required. Of course, Hurricane Sandy is an extreme example, but it proves the point that green web hosting providers should focus on making server reliability one of their biggest priorities, doing whatever possible to guarantee the best possible uptime for sites on their servers.

Customer service

A lack of good customer service went hand in hand in the aforementioned unreliable hosting. In a few cases, it seemed like the entire support staff was one guy poking around to fix issues in his free time, which is entirely possible for a small company. Not only were many of these companies not able to help us when websites went offline, but some also often seemed genuinely disinterested in doing so. At least twice, the responses we received were indifferent and oftentimes terse, their accompanying suggested solutions ill-conceived. One issue took more than two weeks to resolve. Another required an unexpected software add-on to the tune of significant additional monthly expenditures. We couldn't decipher whether these support solutions were outsourced or not.

In some respects, this is understandable—growth can be a significant challenge for any small business. But it is not acceptable. If these companies want to scale—and in order to bring more sustainability to the Internet, we *definitely* want them to scale—they need to have nothing short of top-notch customer service: professional, responsive, and delivered by experts. They must care for their customers as much as they seemingly care for the planet.

A strange twist

In late 2015, we discovered by chance that many of those providers with which we had such terrible experiences over the course of the previous five years were in fact owned by a single, large public company.

Someone shared with me a list from Reddit (*https://www.reddit.com/r/Hosting/wiki/eig*) of web hosts owned by a company called Endurance International Group (EIG). The list contained some very familiar names: green hosting companies with which we had tried to do business over the previous five years. The post read:

> By maintaining dozens of brands which don't clearly indicate they're owned by the same group, EIG is able to retain customers who "switch" from one EIG brand to another. Ultimately, the service customers receive won't change and often times the customer's website is merely moved within the same datacenter.
>
> Due to several complaints regarding downtime, website slowness, quality issues, and horrible customer service, it's become common practice for experienced web masters to warn others to avoid EIG brands. Unfortunately it's not always easy, as EIG intentionally masks their ownership of the brands.

As I skimmed the list, many familiar names popped out at me. I was amazed at how many companies listed were on our own "tried and failed" green hosting list. According to this post, every move to another company (which was inevitably plagued by the same service and reliability issues) was in reality owned by the same parent company. This scenario was all too painfully familiar. Could it be true?

Doing some due diligence to be certain the Reddit post wasn't the isolated frustration of a single angry customer, I came across many other posts from frustrated EIG customers who had logged complaints about quality, reliability, and customer service. In total as of this writing, 345 complaints on the Better Business Bureau's site alone have been logged against EIG.

I sent a query to EIG's PR team requesting more information about their green hosting brands, but received no response. I chose not to list the hosting providers that we explored here (not all of which were part of EIG), as our experience only represents that of one small company. However, I share this story as a cautionary tale for others to do their homework when it comes to choosing a hosting provider that touts its green values. Suffice it to say, we have moved away from hosting with EIG brands.

In the United States and some other countries, public companies are legally required to maximize shareholder value. In other words, shareholders can sue company leaders if they think the company is not doing everything within its power to squeeze all potential profit out of quarterly earnings. In some cases, this can lead to gutting sustainability or corporate social responsibility (CSR) departments, slashing philanthropic efforts, or making decisions that undercut workers, price gouge customers, and do extensive damage to the environment all in the spirit of unlimited growth. Variations on this story have unfolded throughout history. Perhaps a similar version plays out in the boardroom of EIG each time it discusses purchasing a new web hosting company.

Incidentally, this is also the primary reason certified B Corps and Public Benefit Corporation legislation exists: for companies that want to legally consider stakeholder needs alongside those of shareholders. A good green host, like the examples we heard earlier from Green House Data and Canvas Host, will consider all stakeholder needs in its decision-making process. A Public Benefit Corporation that considers these needs is also at less risk of being sued by shareholders for pursuing purpose alongside profit.

Green *and* good: winning big

When you combine the small number of renewable energy–powered web hosting options available with unreliable servers and sometimes questionable customer service practices, it is no surprise that few people have shifted site hosting away from bigger, cheaper, and arguably more stable hosting providers. They offer stability and customer service for an affordable price. Unfortunately, that comes at the expense of the planet.

What the Greenpeace *Clicking Clean* report and our experience highlight is that a huge gap exists between the efforts of behemoth companies to minimize their online impact and what the rest of us who build or otherwise inhabit the Internet are doing to make the Web more sustainable. Mightybytes remains committed to hosting our websites on servers powered by renewable energy, and we're not the only ones. There is a growing number of companies looking for better, more sustainable data solutions.

Andrew Boardman from Manoverboard agrees. "Although we have spent a considerable amount of time and resources on identifying a green host, we have yet to hit the trifecta: a company that is purpose-driven with great customer support and that also happens to be powered by 100% renewables. We won't stop looking. Our business, clients, and planet depend on it. I know that when we get there, we'll all win big."

Green hosting, greener Internet

Although the bigger, consumer-facing players get most of the attention for their efforts transitioning to renewable energy, the rest of the Internet has some serious electricity needs, as well. What about the tens of thousands of smaller companies, nonprofits, and local governments that inhabit the less trafficked regions of the Internet? Most of these companies either rent server space in shared facilities or host their operations with cloud computing vendors and content delivery networks. How can these organizations transition to renewable energy for their data needs?

Though figures on renewable energy use of the smaller colocation companies are elusive, of those audited in Greenpeace's 2015 report—many of which supply server space to the aforementioned types of companies—the average amount of clean energy powering their servers was barely over 14%, which is on par with renewable energy production in the United States. That number is only 13%. So clearly there is significant room for improvement as we move to a clean energy–powered Internet.

That said, progress is being made by many of the larger tech companies that have resources to invest in onsite renewables. AWS announced in its 2015 shareholder letter that the company was on track to power AWS with 40% renewable energy—up from 25% the year before—and had invested in four significant wind and solar farms that will deliver 1.6

million megawatt hours per year of additional renewable energy into the electric grids that supply AWS data centers.[12] Considering that AWS houses as much as one-third of the Internet, that's major.[13]

Finally, even though buying RECs is a small step forward and in some regions the only option, unbundled RECs can do little to improve the supply of electricity directly powering a hosting provider's servers. This is further complicated by a political climate that doesn't support easy access to renewables and a public that is often either indifferent or doesn't understand how to make the transition.

Conclusion

In this chapter, we talked in detail about the importance of web hosting powered by renewable energy and how not all web hosting is created equal. We also covered other components—open source software, frameworks, mission statements, and the like—that can help make the process of creating digital products and services more sustainable with less environmental impact. Of the components described in this chapter, green hosting is arguably the most important, given that it's how our digital products and services get their power. But the other components can help you streamline workflows and ensure that your content is created in an environment where more sustainable choices are encouraged. They can also help make your applications accessible to the widest number of people across platforms and devices.

Action Items

Want to apply the knowledge learned in this chapter to your own work? The following list offers advice on how to move forward:

- Use the Green Web Foundation's Green Hosting database to find several hosting providers that power their servers with renewable energy. Compare notes on features, pricing, and how, specifically, each company commits to renewable energy. Is it worth migrating to one of them? Why or why not?

---

12 Amazon.com, 2015 Shareholder Letter. (*https://www.sec.gov/Archives/edgar/data/1018724/000119312516530910/d168744dex991.htm*)

13 Synergy Research Group, "AWS Market Share Reaches Five-Year High Despite Microsoft Growth Surge", February 2, 2015 (*https://www.srgresearch.com/articles/aws-market-share-reaches-five-year-high-despite-microsoft-growth-surge*)

- If you answered "no" to the preceding question, your existing web host's customer service department and express interest in renewable energy–powered options.

- Explore options for powering your office with renewable energy. Are there local government subsidies in your area for purchasing solar panels? Can you buy renewable power through other means without purchasing unbundled RECs? If not, is there an avenue through which to change this dynamic?

- Does your company have a mission statement or core values that include a commitment to the environment? If not, collaborate to create one.

- Are you using proprietary software? Find an open source equivalent and audit it for feasibility on a current or future project.

- Explore GitHub for potential project resources.

- Commit to developing all your projects with web standards in mind and make sure your team shares this commitment. Find ways to keep each other accountable.

[4]

Content Strategy

What You Will Learn in This Chapter

In this chapter, we will cover:

- Why content "findability" and search engine optimization (SEO) are important for sustainability.
- How you can make your own content more searchable and thus more sustainable.
- Content strategy methods for streamlining the user experience (UX) and making it more compelling.
- A few techniques for making streaming video more efficient.
- Options for executing more sustainable social media strategies.

The Content Conundrum

The content explosion that's happened on the Internet presents a significant sustainability challenge. Facebook's 1.44+ billion global users alone spend an average of 20 minutes or more per day on the site.[1] Users in the United States spend twice that time. That's nearly 30 to 60 *billion* collective minutes per day...just on Facebook! And although the company's servers (as of this writing) are powered by 49% renewables with significant clean energy investments set to increase that number, it is still only one social network—albeit a *really, really* big one—on an Internet that seems to spawn a new social platform every week.

---

1 Jillian D'Onfro, "Here's How Much Time People Spend on Facebook per Day", *Business Insider*, July 8, 2015. (http://www.businessinsider.com/how-much-time-people-spend-on-facebook-per-day-2015-7)

FIGURE 4-1.
Users collectively spend 30 to 60 billion minutes *per day* on Facebook

Similarly, users of popular video platforms like YouTube, Netflix, Hulu, and others make up more than 70% of consumer Internet traffic, a number expected to grow to 82% by 2020.[2] Then there's the growth of bandwidth-heavy rich-media apps like Vine, Instagram, and Snapchat, as well as collaboration apps like Slack or Hipchat, plus virtual reality, IP video-on-demand (VoD), and so on. Newsfeeds alert us every few minutes to what's happening in the world. Add to this hundreds of millions of blogs—some abandoned, some not—that take up space on servers all over the world, each of which use power 24/7/365. And those are just the consumer applications. Internet video surveillance and content delivery network traffic are also exploding.

---

2 *Cisco Visual Networking Index: Forecast and Methodology, 2015–2020* (2014). (*http://www.cisco.com/c/dam/en/us/solutions/collateral/service-provider/visual-networking-index-vni/complete-white-paper-c11-481360.pdf*)

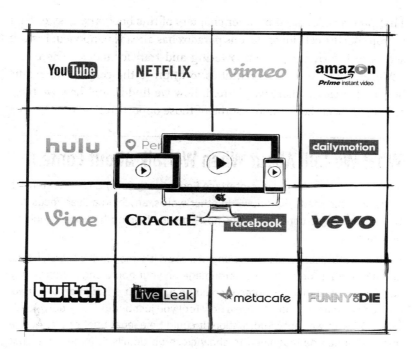

FIGURE 4-2.
Video content makes up more than 70% of consumer Internet data

We have blogged, tweeted, and posted ourselves to a place where content is so ubiquitous that it has the same shelf life as that cup of coffee you got from Starbucks this morning. And much like the cup that coffee came in, as soon as content has served its brief purpose, it is often destined for the great content graveyard of the Internet, out of sight, out of mind. Until, of course, you need to find it again. Then, you search Google, your Twitter feed, your Instapaper account, or even that old warhorse, Delicious, looking to find that one stat that will prove your point in an argument with a friend over cocktails.

All of this leads us to a very important question: How can we create a more sustainable relationship with our content?

Manoverboard's Andrew Boardman says it's quite simple: "All content should be remarkable—and worth sharing. A truly sustainable content strategy is driven by organizations that regularly care for and about their customers. Content should change minds—not drain phones."

That said, as addressed in other chapters of this book, the answer isn't to stop sharing cat videos. Jevons paradox has already happened. Instant access to information is empowering and transformative. There are, however, many great opportunities to improve the content we create, how it is created, where we share it, how we find it, and how we measure its impact. Let's explore some of those opportunities.

What We Talk About When We Talk About Content

Jared Spool, founder of User Interface Engineering (UIE), says, "Content[1] is what the user needs right now." In other words, anything a user needs can be interpreted as content. "Right now" means helping him find it quickly and easily.

In other words, content is *everything*: it's copy, it's images, it's search fields, and it's contact forms. It's the hero image on your home page and the tiny copyright notice in your footer. It could be a product description or a map, a video showing you how to use a new tool you just bought, or a downloadable PDF with warranty information. It could be a legal brief or a resizable image that can be magnified to show close-up details. It might be a chat window that gives you real-time access to customer service, a game, or perhaps an object you can interact with in 3D. Content is all these things... and more.

How quickly you get users in front of the content they need at any given moment is what constitutes a more sustainable content strategy.

---

1 Source: https://medium.com/uie-brain-sparks/content-and-design-are-inseparable-work-partners-5e1450ac5bba#.3fm94eidn

---

TOWARD MORE SUSTAINABLE CONTENT

Author and content strategist Kristina Halvorson seems to have the de facto definition of content strategy in her aptly titled book *Content Strategy for the Web* (New Riders, 2012):

> Planning for the creation, delivery, and governance of useful, usable content.

If while implementing content strategy, we also plan to meet present needs without compromising those of the future, as a more sustainable philosophy would dictate, how do we proceed? In this glut of online information, how do we devise content strategies that meet our own organizational needs while also considering their environmental impact?

Chicago B Corp Orbit Media's Andy Crestodina, author of *Content Chemistry: An Illustrated Handbook for Content Marketing* (Orbit Media Studios, 2014), says there is a blend of interests between sustainability and content marketing goals. "A content marketer wants lots of visits, and we want those visitors to stay as long as possible," he says. "We want low bounce rates and long visits. Those are considered success metrics. But in the long run, we need to think about the desired outcome for us and for our visitors, and look for ways to meet those goals as efficiently as possible."

Andy suggests asking yourself a couple of important questions:

- Are your users finding what they're looking for quickly? Or, are they bouncing back and forth between pages looking for more information? If so, the bounce rate might be low, and the time on site might be high, but you're not meeting your goals or theirs. That's a problem.

- Are you attracting the right visitors? Ranking for an irrelevant phrase or inviting visitors who aren't likely to act might look good for your traffic trend line, but again, this doesn't meet your goals or theirs.

When considering sustainability alongside content strategy, it is critical to think carefully about what success really looks like. "You may find that some of the metrics that look nice in Analytics are just burning processor power and heating up a data center," Andy says.

To understand how we can think more sustainably about our content, we first must be very clear about our own goals and how to meet them as efficiently as possible. Oftentimes, the elements that make a product or service more sustainable are those that also make it good for users. But if you're not paying attention to the right details, it can be easy to fall into a trap of producing content that no one cares about without realizing it.

Content and UX: Partners in Crime

Author Karen McGrane tells us that working on design without also considering content is like giving someone a beautifully wrapped empty box for their birthday. Although content and design are addressed in this book in separate chapters, it is important to note that tasks associated with each discipline are often best executed alongside each other in collaborative settings. A good design process will address content needs with UX tasks and require that content types and patterns be defined before starting exercises like wire-framing or component design. Likewise, content strategists worth their salt don't work in a vacuum: the most successful content strategy is one created collaboratively with relevant stakeholders and is developed hand in hand with UX.

Working these disciplines in tandem accomplishes several important things:

- Rather than considering it as an afterthought, this approach establishes content as a key component in successfully executing digital products and services
- It lets teams maintain a "content first" approach to their projects, which is crucial when establishing display patterns across devices and platforms.
- Using several types of real content in the design process can help teams quickly identify scalability problems much more easily than placeholder images or *lorem ipsum* text will.

DEFINING THE RULES

So what comprises a more sustainable content strategy? The process itself shouldn't have a lot of waste and the results should be actionable and clear to users. First, the workflow for creating content should be efficient and collaborative in a way that gets to desired outcomes quickly. The content created should be easily found, organized clearly, tell a compelling story, and most importantly, motivate users to take action of some sort. It should also use display patterns that are relevant to the devices and platforms upon which it is being viewed. Finally, the road to more sustainable content begins with measuring the important stuff. Doing more of what works and less of what doesn't is the quickest way to greening up your content's "supply chain."

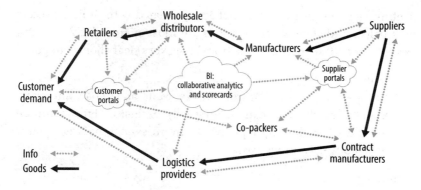

FIGURE 4-3.
Can your content have a supply chain? If so, what does it look like?

Measurement is clutch

You can't track what you don't measure. If you're not measuring performance, you don't know what's most useful. If you don't know what's most valuable to your users, it will be more challenging to trim waste, streamline your process, or create new content to meet their specific needs. You can measure most anything on a website or mobile app but that doesn't necessarily mean you should. But where to begin? Where can you find the data to measure? Here are some ideas:

Analytics
> Metrics platforms like Moz, Hubspot, or Google Analytics will tell you which content is working and which isn't. But be careful: just because these platforms can measure everything doesn't mean you should. It can be easy to waste a lot of time measuring things that don't matter.

Tree testing
> This is when you test navigation structures to help users find content more quickly.

A/B testing
> This is when you provide two different content or design options for users and track which performs better.

Conversion optimization
> This is when you test and improve a page or screen's ability to incite users to perform desired actions.

Defining the most important metrics associated with project success will help you dictate which methods are most appropriate for testing results. And those metrics come from being very clear about your goals:

Define needs
> Be specific about what your users want and what you want. Sometimes the two can be mutually exclusive. Find common ground where both groups' needs are in sync; otherwise, you're just spinning your wheels.

Define a goal
> It is best to be specific, such as "grow email newsletter subscriptions by 10% in the first quarter."

Define content types
> Which content will most help you reach those goals? Is it blog posts with clear calls-to-action (CTAs)? Downloadable white papers? All of the above? Plan to test each. Use those that work best.

Associate analytics
> What metric will prove that goals are being met? In the preceding example, it could be tracking white paper downloads or month-by-month subscription sign-ups.

Benchmark it
> Show improvement over time and compare it to your original goal. Are you reaching your goals consistently each month? Reevaluate and iterate. Repeat this cycle as necessary.

We will cover variations on this in the section "Agile Content: A More Sustainable Solution?" on page 129.

The content audit

With so much industry emphasis on *creating* content, how do you develop good practices for phasing out content? How do you figure out what's working and what isn't? Who will clean up the inevitable mess left by abandoned or nonperforming pages? Knowing this will help you to keep sites lean and streamline your content strategy moving forward. A great place to begin is with a content audit. It will help you to define the content that performs best and identify the weak spots.

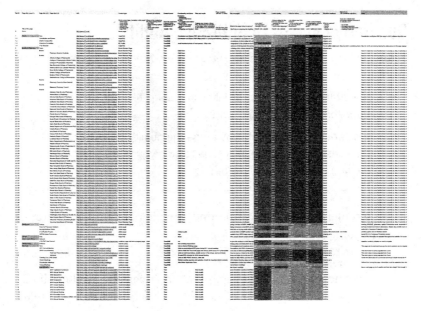

FIGURE 4-4.
A content audit will help you to figure out what's working and what's not (yes, it's a spreadsheet)

A content audit contains a list of all the URLs on your site and defines important metrics like who owns the content; how it performs; if the URL is going to be retired, where should it redirect to; and so on. Some teams include keyword research in their content audits, as well. They can help content teams assess content performance, identify trends and content gaps, and keep track of editorial changes. Content audits are especially useful in the process of a redesign (because they require that a certain amount of content already exists). They are also action-driven. Here are some common action items when creating a content audit:

- Remove (Can you dump it?)
- Keep as is (Don't touch it.)
- Improve (Can you make it better? If so, how?)
- Consolidate (Can it be merged with other content? If so, how?)

Qualitative Versus Quantitative Audits

Before diving into a content audit, first figure out which type of audit you need:

- A *quantitative audit* is a comprehensive, soup-to-nuts record of all your content, including all URLs.

- A *qualitative audit* can help you judge the quality of your content against a specific set of criteria to help you make plans for improving individual sections or pages.

Keeping content sustainability in mind, consider two important things during a content audit:

Cut ruthlessly
> There's no sense keeping pages that don't perform well (or at all). If a page isn't one of your top x performers (you pick the number relevant to size and scope of your project), it should go.

Redirect pages
> When cutting pages from a site, be sure to redirect the original URLs to the most closely relevant pages that still exist. This will keep search engines from registering 404 errors on your site, and it will help users get to the next best thing from what they were looking for.

In content marketing, failed efforts happen all the time. Iteration is key: you try a bunch of things and then do more of those that work, less of those that don't. It's a pretty simple equation. The problem is that most of the failed efforts tend to remain online, despite the fact that they serve little purpose. Outdated, unused, cryptic, or hard-to-find content is just taking up unnecessary space on your web server. It takes electricity and disk space to house that data.

But aren't pages that get visited most often the ones that use the most electricity? Absolutely. If those pages are optimized properly, they are serving your company or organization's needs as well as those of your customers, donors, and so on. Hence, they are helping to sustain your business. And that's not a bad thing.

Overall, good measurement practices can help you pick and choose which content should stay and which should either be improved or deleted altogether, and a content audit can help with that. If you want to make an existing digital product or service more sustainable, begin by auditing your existing content. Here is a quick content audit experiment to try:

1. Open Google Analytics.

2. Run a report on the 100 most-visited pages over the past year.

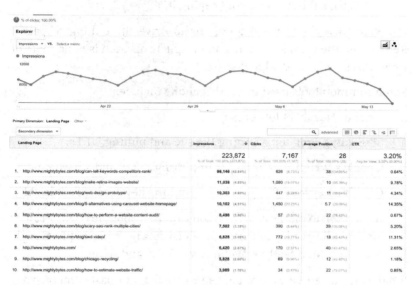

FIGURE 4-5.
Should I stay or should I go now? Google Analytics can help you answer this question

3. Now, ask yourself these questions:

   - Are the pages listed at the top those that you want people to visit most?

   - Are people taking actions on those pages? In other words, are they serving your business needs? If you don't know, you might need to set up some goals and funnels.

   - Can you improve those pages to better serve their intended purpose and drive conversions?

- Do you see any surprises? Are there pages that should be in the top ten that are not? Are there pages at the bottom of the list that are key to your organization's success?

4. Now look at the bottom of the list, those URLs that live on the island of misfit web pages. Ask yourself these questions:
   - Why is that page even a thing that exists?
   - Why did it fail?
   - What's the rationale for keeping it?

Use your answers to these questions to drive the culling of pages. Remember, the most sustainable (and Agile) approach is to do more of what works and less of what doesn't.

Some common tools used in content audits include:

- Excel or Numbers (of course).
- Screaming Frog for scanning the site and pulling URLs.
- Google's Webmaster Tools for importing query and keyword data.
- Google's Keyword Planner for researching keywords (if applicable).
- If you want to merge data from some of these tools, use URLProfiler for compiling links, content, and social data from places like Google Analytics, Moz Metrics, Webmaster Tools, and so on.

Final word: it can be easy to get sentimental about past content. Your boss or the company's CEO might want those photos of their fishing trip on the blog. If they're not converting, they're just burning pixels. If no one's visiting, they're just wasting server space. Ditch them. A content audit can help you make the case for that.

Information architecture

Content must be organized in a way that helps users find what they need as quickly as possible to maximize efficiency and be considered a more sustainable choice. This can pose challenges to information architecture.

For example, when web copywriters or brand strategists create clever or pithy button labels in the name of being "on-brand" this can significantly disrupt the ability for users to find what they need. These cryptic labels can also confuse users with disabilities who may need enabling technologies, such as screen readers, to experience your content.

FIGURE 4-6.
Is it really awesome? Be clear with action words on buttons

Worse, structural issues and language clarity can go unresolved due to lack of clear ownership because information architecture questions often arise periodically throughout the life cycle of a project and thus affect all parties. Information architecture can be one of those nebulous terms where responsibility may not clearly be dictated. Is it the content strategist's purview? The UX team? Decisions made about information architecture affect everyone on a project—most of all, users—so it's important that conversations about navigational structure and labeling happen early and often. Design teams should clearly designate someone as "keeper of the IA" early in a project so that when these issues inevitably arise, it's someone's responsibility to address them.

To make content easier for users to find, test your navigation labels and drive information architecture decisions with test data. Should the navigation section be labeled "Portfolio" or "Work"? Should the form submission button say "Submit" or "Grade Your Work"? Is it easy for users to find content when given a specific task? These are important questions to which you need to know the answers. If you fail to answer them successfully, users will simply leave. Or worse yet, they'll become frustrated through a nightmare of trial-and-error processes.

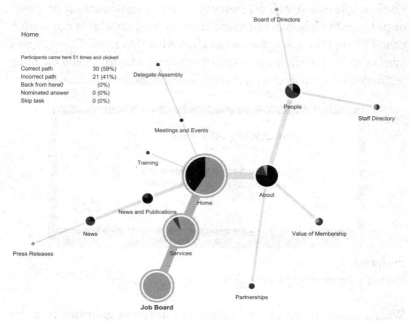

FIGURE 4-7.
When tasked with finding something specific on your site, how many users will be successful on the first try?

Tools like Treejack from Optimal Workshop, UsabilityHub's Nav Flow Test, or UserZoom can be infinitely helpful to drive your navigation decisions with data. Users are given specific tasks like "make a donation" or "find a book by George Orwell," and then they are faced with navigation structures planned for a digital product or service. If the majority of users successfully complete the tasks assigned, you're in good shape. If not, you'd better change it up a bit.

These tools give you helpful metrics such as percentage of completion, time to completion, how many completed the task in the first try, and so on. This is incredibly helpful data when trying to streamline the amount of time it takes a user to find what they need.

STORY MATTERS

It can be easy to get caught up in all the metrics, the navigation labeling, and the thousand other choices you need to make to ensure that your content performs well. But let's not forget that when creating content, good storytelling is critical. After all, getting users engaged and inciting them to take action quickly is really what the Internet is all about. It's also difficult work.

"Thankfully, writing is not one-way communication," says Jill Pollack, owner of Chicago B Corp StoryStudio Chicago, a community and training facility for creative writing and corporate communications. "Good writing invites the reader to get involved in a dialogue and I believe that's what makes good content marketing, too. Make sure you offer ways for your reader to participate in the discussion."

"For anyone who employs content marketing, storytelling is key. It's how humans have always learned and how they have always built community. Good stories come from having an authentic voice and the sincerity to build trust."

Often details are what make a good story. On the Internet, content creators walk a thin line between sharing too many details and not sharing enough. Too many details and you risk losing the audience; not enough and, well, same thing...but for a different reason. How do you know you are sharing the right amount of details, or for that matter, telling a compelling enough story?

FIGURE 4-8.
In workshops at StoryStudio Chicago, Jill helps people find just the right amount of details to tell a compelling story

"My best advice is that whatever you are writing or sharing with your audience, take your time with it," Jill says. "Read it at least three times. Consider the message. Have someone act as your editor to proofread and challenge your assumptions. And then, after all that, read what you have written and ask yourself if you have reached high enough and whether you've asked your audience to dream big enough. The biggest mistake content marketers can make is to underestimate their audiences."

Here are some quick guidelines to remember:

- Be clear: readers appreciate clarity.
- Share your value proposition right up front.
- Get to the point, but don't leave out important details.
- Always end with an ask or action item. Always.

But what about search engines? Isn't this chapter about SEO and quickly finding content? It is indeed. But Andy Crestodina says good story telling and search engines aren't mutually exclusive, nor should search strategies trump content quality. Ever. Andy recommends following these steps to align your content with search marketing efforts. Don't target a phrase or promote a post in organic search, he says, unless you first do the following:

- Sincerely attempt to create the best page on the Internet for your topic
- Target a keyphrase only if it is truly relevant to your content
- Use natural language, with full sentences (questions and answers) to help Google present simple bits of data right in search results, if possible

Although we could spend a great deal of time in this chapter talking about search algorithms, disavows, "black hat" SEO practices, and keyword ranking, Andy and Jill make the most relevant points when it comes to your content. Be true to its purpose and intent first.

Leading by example

Finally, there is one more very important role that good content can play: it can help users make more sustainable choices. Whether that means highlighting a more sustainable shipping option for a purchased product, rewarding users for making more conscientious choices, or simply by telling a compelling story about why one option is more eco-friendly than another, good content can make a real difference. Always be on the lookout for opportunities to highlight sustainability first. Don't believe me? Test it and tell me what you find out.

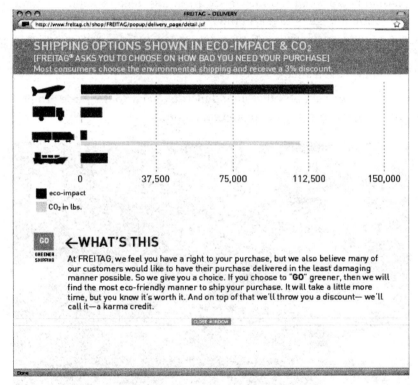

FIGURE 4-9.
Great content coupled with good design can help users make more sustainable choices

VIDEO CONTENT

Using video to tell a story can be a more compelling way to get a message across. It can also be more bandwidth-intensive and use more energy. With video content taking up the majority of consumer Internet traffic, it's important that videos get to the point, are compressed properly, and when possible hosted on servers powered by renewable energy.

The story diet

Andy Crestodina notes that putting your script on a diet is the first place to begin when it comes to streamlining video content. "Skip the long intro and get straight to the point," he says. "If you can get your five-minute video down to three minutes, you just did everyone a favor. With video making up so much of online data, this is good for both the viewer and the planet."

In other words, a web video should increase value to the same extent it uses more energy to produce and consume.

FIGURE 4-10.
Video shoots always capture more footage than needed. Streamlining your script can help ensure that's not an overtly wasteful process.

Content guidelines for video scripts are very similar to those mentioned before but even more important in the editing process. Video shoots always capture more content than they need and it can be very tempting to save time by just leaving this bit or that bit in your file. Don't do it. If it doesn't fit your video's specific purpose, if it doesn't serve the goal of informing, compelling, or converting users, get rid of it.

This is doubly important for videos that play on mobile devices, which at this point is all of them. Even though your video platform of choice will do its best to serve files at a bitrate that's appropriate for your bandwidth and processor speed, a seven-minute video that should have been three still wastes four minutes of a user's time, four minutes' worth of streaming data, and four minutes of battery time on the user's device. That's bad news all around.

Video workflows

Streamlining the video creation process can make for a more sustainable output. The tools used to create and edit video can use up large amounts of electricity. The same principles we covered in our discussion of sustainable workflows in Chapter 3 apply here as well.

Canadian media production company and B Corp Hemmings House creates documentaries and educational, promotional, corporate, and many other types of video-based content. CEO and Executive Producer Greg Hemmings says that sustainability is very important to the company. "Not only does sustainability reference the health of our environment, but also our communities, our society, and our business model," he says. The Hemmings House team regularly explores new ways in which to apply the sustainability principles they apply to the rest of their business to their digital footprint. "Hemmings House is a digital firm. We still have a ways to go in reducing our digital carbon footprint, as video production produces very large files. But we're getting there"

In addition to encouraging employees to bike and walk to work, supporting local businesses and sourcing local as much as possible, Hemmings House takes additional steps to make its production workflow more sustainable, as discussed in the following sections.

FIGURE 4-11.
Hemmings House is always exploring new ways to make production workflows more sustainable

Development/preproduction. When preparing for and in the early stages of a project, Hemmings House uses the following tactics:

- Highly recyclable paper used in the office.
- For larger print-related projects, use other B Corps such as Mills Office in British Columbia.
- Digital call sheets.
- Production meetings are handled with Skype or Google Hangouts to avoid in-person development meetings unless absolutely necessary.
- Employees walk and bike to work as much as possible.
- The company belongs to Sustainable St. John, a local membership-based sustainability organization.

Post-production. In the office and edit suites, the Hemmings House team does the following:

- Implements low-energy consuming standards (turning computers off at the end of the day rather than running them all night).
- Work in sunlight when possible rather than using overhead lights.
- Turn the heat down at the end of the day.

FIGURE 4-12.
Hemmings House tests project files on mobile devices to ensure file sizes are small but maintain quality

Their editors use a significant amount of electricity with power-hungry editing apps like Avid, Photoshop, and After Effects. They reduce consumption by always setting computers in power-saving modes, using sleep modes, and shutting down all computers when not in use.

They also transfer a number of files via transfer services like Hightail. "We believe this is less of an environmental impact than mailing footage on a USB key or hard drive," says Greg. "However, finding a green-powered FTP service is something we should do going forward."

Distribution. Green hosting with servers powered by renewable energy is the next step for Hemmings House in addressing how the company distributes video files. It currently hosts most of its videos on YouTube and Vimeo.

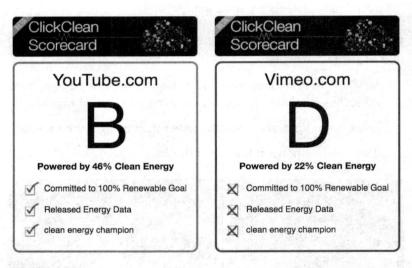

FIGURE 4-13.
Where to host your video: although YouTube, as part of the Google family of products, is carbon neutral, Vimeo gets low marks for environmental sustainability from Greenpeace

Video compression

The best way to keep bandwidth use down is to encode video with the highest possible compression while still maintaining image quality. Streaming services will automatically compress the video to many levels of quality and then dynamically swap in the best possible version that a viewer connection can support. YouTube aspires to streaming in HD if that's what you uploaded. But is it really necessary? For many videos, it's not.

The audio tracks of your video files is another area where you can save on bandwidth. Talking head videos don't need stereo audio tracks. "Export the audio as mono," says Andy Crestodina. "Save some bandwidth and your viewers will never notice."

Accessible video

Finally, it is important to make video and audio accessible to users with disabilities. The W3C lists the following recommendations for making a video accessible:

- Text-based transcript of the script. This is required for accessibility compliance, but it's great for search engines, too.

- Audio description that explains what's in the visuals. This is required only for content where audio does not already describe what is in the visuals, such as in things like charts and diagrams.

- Onscreen captions are important when people need to see what is happening in the video and get audio information at the same time.

- Sign language is helpful for hearing-impaired people who use signing as their primary language. This is not typically required.

AGILE CONTENT: A MORE SUSTAINABLE SOLUTION?

Lean, Agile, growth hacking, Scrum, Kanban. These variations all have a common goal: make better decisions faster. The results come in different forms, but they often use fewer resources while yielding higher quality outcomes, which could be considered as more sustainable (each project being its own unique snowflake, of course). As we discussed in "Measurement is clutch" on page 113, the best way to use fewer resources is to do more of what works and less of what doesn't. The best way to do that is by measuring what you learn each step of the way.

Agile methods support this philosophy, and although they are most commonly applied to software projects, they are really relevant to nearly any project or initiative. Author Pamela Meyer shows how they can be applied to entire organizations in her book *The Agility Shift: Creating Agile and Effective Leaders, Teams, and Organizations*.[3]

"The success of any project or process revolves around the ability to effectively communicate, collaborate and coordinate, as well as to learn continuously," says Pamela. "Leaders and teams in organizations across industries are making agility a strategic priority and shifting their mindset and practices to stay competitive in a rapidly changing market."

Being agile in your content means people need the authority to run experiments. Teams should have the knowledge to track important metrics and make decisions to pivot (or not) based on what they learn. They should be able to react quickly both to external forces, like an industry upset or new development by a competitor, and to internal forces, such as new company developments, budget cuts, customer service needs, and so on.

For many companies, some version of this process might likely exist already. But here's one thing that almost always is overlooked: organizations also need processes in place to clean up all that junk that didn't work. It takes up server space and wastes electricity. Good content teams also have the ability to regroup and do housecleaning to reduce waste when necessary.

A scientific approach

We discuss ongoing testing with real users as part of any project in other sections of this book. The knowledge gleaned from running small content experiments can yield insights that inform how best to proceed to the next step, reducing necessary resources to execute. Creating a content hypothesis and then proving or disproving that hypothesis can greatly help reduce wasted efforts. Even though it can be very tempting to give in to the content beast, publishing and posting everything that

---

3 Pamela Meyer, *The Agility Shift: Creating Agile and Effective Leaders, Teams, and Organizations* (Brookline, MA: Bibliomotion, 2015).

comes to mind, it is best to take a scientific approach to your content. "Content marketing must be approached like a scientist.[4] There are no mistakes, just feedback," says Search Engine Journal's Srinivas Rao.

Lean startups and most Agile organizations accomplish tasks in iterative cycles with very specific target goals and measurable outcomes. Taking a tip from them could yield potentially more sustainable content strategy and marketing results. At the very least, the process will be more sustainable because its very nature is built around reducing waste.

In its simplest form, a content experiment could go like this:

1. Specify a goal.
2. Specify a metric—or key performance indicator (KPI)—that represents the goal.
3. Create content to achieve the goal.
4. Analyze the metric—did you get closer to the goal?
5. Improve and try again.

Look familiar? Let's apply that to a common sales funnel scenario for an online subscription software company trying to increase sales. A 2014 study by BrightLocal on how online reviews and testimonials influence purchasing decisions showed that nearly 9 out of 10 consumers have used online reviews to help them make decisions about whether to support a local business (4 out of 10 do so regularly).[5] In this scenario, the company hypothesizes that it can increase sales 10% by adding customer testimonials to its website. To reach this goal, it knows that it needs to sell an additional 100 subscriptions in month 1, 110 in month 2, and so on, and so on (keeping the math easy here).

The team collects several quotes and photos from happy customers. It then adds short customer case studies to key sections of its website with CTAs. It configures goals and funnels in Google Analytics that

---

4 Srinivas Rao, "Applying Principles of Growth Hacking to Content Marketing", Search Engine Journal, October 6, 2013. (*http://www.searchenginejournal.com/applying-principles-growth-hacking-content-marketing*)

5 BrightLocal, "Local Consumer Review Survey 2014". (*https://www.brightlocal.com/learn/local-consumer-review-survey-2014*)

track users from the aforementioned CTAs on each case study page through the process of a purchase. If users get to the "Thank you for your purchase" page, a goal has been met. Then, the team tracks performance of that goal for three months.

Here's how that might break down:

1. Goal = increase sales lead 10% month-to-month.
2. KPI = # of completed goals in Google Analytics.
3. Content = featured customer case study.
4. Analysis = Reached 60% of goal in month 1, or 60 additional subscriptions.
5. Improve content and run experiment again.
6. Analysis = Reached 88% of goal in month 2.
7. Improve content and run experiment again.
8. Analysis = Reached 112% of goal in month 3.
9. Improve content and run experiment again.

In the preceding example, though the experiment initially came short of the goal, results were promising enough that it was worth running the experiment again with some tweaks. Improving case study content might increase conversions, for example, or perhaps the CTA wasn't clear enough.

But what if the test had yielded different results? What if, after running the experiment several times, they still didn't reach a desired goal; perhaps it's time to try another tactic with a different type of content.

It is also important to isolate whether the users who converted actually came from the efforts of your experiment. In this example, it would be critical that the first step in funnel measurement started on the case study page. Even though the end result might be the same as in other experiments, it is important to segment the traffic in a unique funnel that presents you with just the information you need to know.

The important thing to remember is that this process was devised from the ground up to reduce waste. If something is not working, the data tells you when to pivot so that you to waste less energy by concentrating your efforts on things that do work.

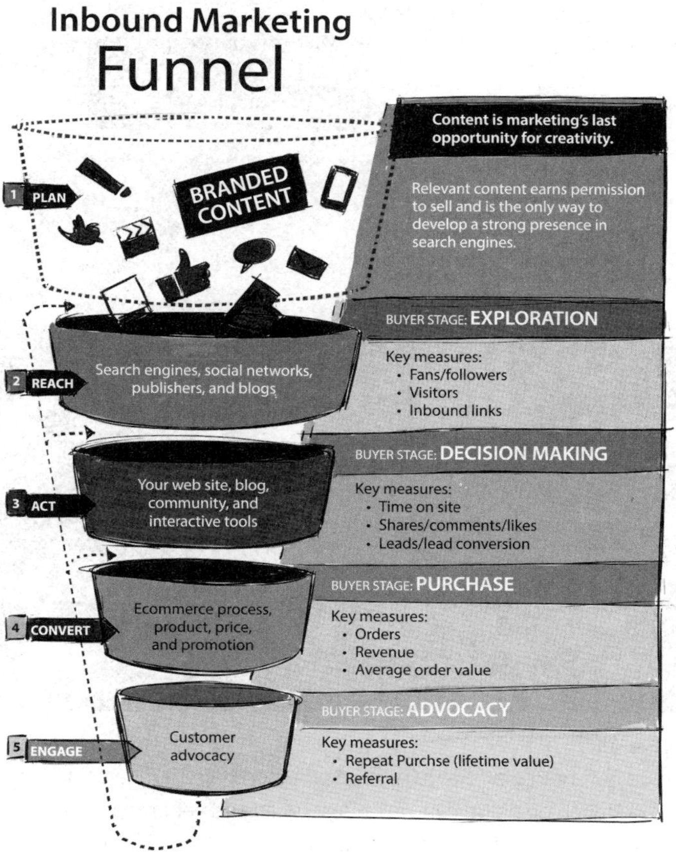

Inbound Marketing Funnel

1 PLAN — BRANDED CONTENT
Content is marketing's last opportunity for creativity.
Relevant content earns permission to sell and is the only way to develop a strong presence in search engines.

2 REACH — Search engines, social networks, publishers, and blogs
BUYER STAGE: **EXPLORATION**
Key measures:
- Fans/followers
- Visitors
- Inbound links

3 ACT — Your web site, blog, community, and interactive tools
BUYER STAGE: **DECISION MAKING**
Key measures:
- Time on site
- Shares/comments/likes
- Leads/lead conversion

4 CONVERT — Ecommerce process, product, price, and promotion
BUYER STAGE: **PURCHASE**
Key measures:
- Orders
- Revenue
- Average order value

5 ENGAGE — Customer advocacy
BUYER STAGE: **ADVOCACY**
Key measures:
- Repeat Purchse (lifetime value)
- Referral

FIGURE 4-14.
A marketing funnel can provide helpful guidelines for running content experiments

Preparing over planning

Planning is wonderful and necessary. It is human nature to plan. Too much planning, however, can use up unnecessary resources. Those resources are wasted if the idea isn't validated by the people who are actually going to use the content, software, or application (or whatever your end product happens to be). One of the many great things about taking an Agile approach to content is that you can prepare versus plan. In other words, rather than meticulously planning out all your content efforts, focus on preparing to use what you learn from each content experiment to inform efforts on the next iteration.

FIGURE 4-15.
It's human nature to plan things, but unexpected twists lurk around every corner. Too much planning can waste resources and set unreasonable expectations; it's better to be prepared.

We once waited two months for a set of personas from a content strategist. Were there quick experiments we could have run during those two months that might have yielded similar, actionable insights? Of course there were.

This is not to say that you should throw planning entirely out the window. Just make sure it's the right amount of planning. Although an editorial calendar can be helpful to keep content marketing teams on track, for instance, it can also be limiting. If something lives on your editorial calendar because it's always been there, yet the results don't perform well, that calendar is doing you a disservice. Similarly, spending exorbitant amount of time on strategic planning can be detrimental if the strategy in said plan doesn't actually work.

Emily Lonigro Boylan of LimeRed shares this example:

> We were waiting for a client to finish up a six-month strategy project. We couldn't build their site until this strategic effort was over. Or could we? Maybe we could help, maybe there was a component of the site we could build and launch that would raise their donation rates right away. Maybe something in the six-month strategy would be outdated by the time it was over. Embracing agile methods can fix that. When you think of what you can save in personnel time, file creation, revisions, wasted effort, it's staggering. If your project can support rapid iteration, why not?

Lessons from Agile content strategy

Here are some lessons learned from applying these methods to content strategy and marketing efforts:

Embrace limitations

Sometimes, the simplest solution can be right in front of you and you don't even realize it. Embrace the limitations of your timeline, budget, or resources and try to find simpler, faster tasks that might accomplish the same results. For instance:

- *Start small*
  Use social networks to bounce ideas first before executing them.

- *Set realistic goals*
  Keep word count low, one idea per post, and so on.

- *Spread it thin*
  Create something small—like a prototype—that you can share with a small but targeted group of people for feedback.

For example, when we considered starting a blog series on sustainable web design practices, we created a single blog post outlining goals for the series. We then tested that post with a targeted network of agencies, B Corps, Climate Ride beneficiaries, and sustainability practitioners. The feedback we received inspired us to proceed. Social media can provide an effective channel for this approach as well.

Always test
This should be pretty obvious by now, but here are some ways to keep testing an important and ongoing part of your process:

- *Create a fast feedback loop*
  Use digital tools to get feedback quickly, early on.

- *On-the-street interviews*
  Ask quick questions of random strangers.

- *Find real users*
  The importance of feedback from real users cannot be overstated.

- *Keep them involved*
  Figure out ways to incentivize real users if that's possible.

On many projects, we have used SurveyMonkey and other simple digital tools to quickly validate our ideas, when possible with real users. It is amazing how helpful people will be if you just ask nicely. Plus, as just noted, with content it can be easy to use social networks, online groups, or UX communities like UsabilityHub.

Measure what matters
Alistair Croll and Benjamin Yoskovitz, who wrote the book *Lean Analytics* (O'Reilly Media, 2013), suggest finding that one metric that matters and focusing on it. That one metric might change over time, but by only focusing on the most important thing, you can be assured that your efforts don't become splintered. Keep the following guidelines in mind:

- *Choose wisely*
  Don't become overwhelmed by metrics.

- *Benchmark it*
  After you have defined it, iterate often until you discover what really works.

For example, when we began building our website sustainability tool, Ecograder (which we will cover in more detail in Chapter 7), building awareness was our one key metric, so we focused on top-of-funnel metrics, such as likes, mentions, retweets, shares, and referral traffic from social networks, and tried to increase those.

Consider curation
> Even though this might not work for every endeavor, curation can be an easy way to find out how something performs without investing significant efforts in it. Consider using the following approaches:
>
> - *Share*
>   By sharing similar content created by others first, you can get an idea of how yours might perform.
> - *Listicles*
>   Create a list of other posts about a certain strategic topic. Test performance of the post.
>
> Content shared by other people is how I learned about sustainable web design techniques in the first place. Similarly, many early posts on the topic were curated lists of other people's posts and additional resources. These can be very helpful ways to quickly gauge how a piece of content might perform.

Repurpose your content
> Repurposing your content into new forms can help you save resources. Don't simply duplicate your existing content though. Put some time into making each piece unique, though based on the same concepts. Because search engines can't identify which version of a piece of content is the original unless you inform them so, you could suffer search ranking penalties and traffic losses.

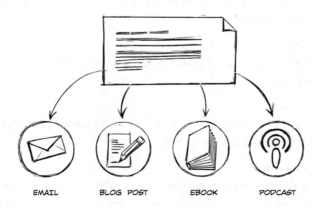

FIGURE 4-16.
An email is a blog post is an ebook is a podcast—content is everywhere, so repurposing it can help you save time and resources

If you do get requests to share a post you wrote on another blog, use a "rel=canonical" tag in the page's HTML head that indicates to search engines that your first post is the original. Consider using the following approaches:

- *Content types*
  Translate blog posts to ebooks, videos to podcasts, and so on.

- *Guest posts and social networks*
  Rewrite blog posts to use on external sites or social networks.

We used content from our blog series on sustainable web design to create a sustainable product development manifesto, which is a downloadable resource meant to help product teams make more sustainable choices when developing a digital product. We also used many of those blog posts as the basis for presentations, other guest posts, videos, and so on. Guest posts on high-ranking sites will also help you build high-quality inbound links to your site.

Drive conversions

A/B testing content can be an effective way to determine which version works best quickly. Sure, it takes more time to create both A & B variants, but you can potentially glean more valuable info to save time on future efforts. If possible, test only one thing at a time, like language used for a CTA or placement of a button. Testing multiple items at once—known as *multivariate testing*—can lead to muddled results because it can be difficult to isolate which specific item caused the test winner to perform better. Consider using the following strategy:

1. *Create two versions*
   Figure out what you want to test and create two variations of your page.

2. *Run a test*
   Use a tool like Optimizely or UsabilityHub to split test two versions of the same page.

3. *Track results*
   Whichever version performs noticeably better is the one to choose moving forward. Use this information to inform future efforts, as well.

Old **A/B Test Winner**

Labeling counts!
10% increase in conversions

FIGURE 4-17.
It's a draw: A/B testing can help you to quickly figure out the best approach for page content items

When creating Ecograder, we knew we wanted to collect the email addresses of those who used the product, but our team had differing opinions on how to accomplish this. Some wanted the email collection to be mandatory. Others thought that was intrusive. So we split-test some options over the course of a couple weeks. In the end, we came to consensus on a prechecked but optional box that added email addresses to our database. This decision was based on which options in our tests performed best while still giving us what we needed (which was to collect emails).

MORE SUSTAINABLE SEARCH

Content can't serve its intended purpose if it can't be found. We have all been there: you need a piece of information quickly, so you pull out your phone, type some keywords into a search engine and are rewarded with 3.76 million results, only a few of which offer the information you need. You spend a few minutes bouncing from result to result not getting the right answer—rarely moving to page two of the results unless you have to—perhaps instead refining your criteria and running another search. All the while, you're burning electricity: on the frontend as you pore over results; on the backend as you run multiple searches. But should you really care about that?

Back in 2009, Google noted that a typical search is equivalent to about 0.2 grams of CO_2.[6] By its own admission, "the average car driven for one kilometer produces as many greenhouse gases as a thousand Google searches." So the impact of a single search is minuscule. But consider that, according to Internet Live Stats (as of this writing), Google processes about 1.2 *trillion* searches per year—the Jevons paradox in action once again.[7] That's still potentially 240,000 annual metric tons of CO_2 generated by one search engine, which is nowhere near the impact of the more than 1.2 billion vehicles on the road—each emitting about 4 tons of CO_2 annually—but still significant.[8] When we talk about minimizing greenhouse gases, every little bit helps.

Now, Google claims that it's been carbon-neutral since 2007, so presumably the CO_2 produced by all Google searches has been either offset or powered by renewable energy, which is a great thing.[9] What Google's numbers don't likely take into account is all that time you spend bouncing from search result to search result trying to find the information you need. That burns electricity, too. A lot of it.

The philosophy here for sustainability is the same as in Chapter 3: the easier your content is to find, the less electricity is burned on the frontend. Although sorting and categorizing all the Internet's websites might be Google's problem, making sure people can find your own content is yours. Search marketing (including SEO) and social media marketing are two ways by which you can get your content in front of the right people at the right time.

But it only begins at *finding* content. After that content is discovered, it's equally important that your content be compelling, answer questions quickly without barriers, and help people make more sustainable choices. In other words, if someone wants to purchase a water bottle from your site, for example, how can you help them make clear choices quickly and incentivize them to choose a less impactful shipping

---

6 Urs Hölzle, "Powering a Google Search", Google Official Blog, January 11, 2009. (*https://googleblog.blogspot.com/2009/01/powering-google-search.html*)

7 Internet Live Stats, "Google Search Statistics". (*http://www.internetlivestats.com/google-search-statistics*)

8 John Voelcker, "1.2 Billion Vehicles On World's Roads Now, 2 Billion By 2035: Report", Green Car Reports, July 29, 2014. (*http://www.greencarreports.com/news/1093560_1-2-billion-vehicles-on-worlds-roads-now-2-billion-by-2035-report*)

9 Google, "Google Green". (*http://www.google.com/green*)

option? Or what if they have general questions that could easily be answered by an FAQ page? Good content practices—strategy, information architecture, optimization, measurement, and so on—can help.

SEO and sustainability

When I asked him about SEO and sustainability, Andy Crestodina noted that saving visitors time and saving data center energy are one and the same. "True SEO is about cooperating with a search engine to help people get to the right answer quickly and efficiently," Andy says. "The goal is to create a page with a complete, detailed answer that ranks for a relevant phrase. This means visitors who find the page are likely to stay, find answers to their questions, and complete their goal. It's inefficient, both for energy and time, to try ranking a page that doesn't answer the question completely."

So, does that mean search engine-conscious content creators shouldn't use keyphrases anymore? No, but it does mean that they're not the only thing to consider. First and foremost, as Andy said, focus on quality. Many other factors—such as social media, user reviews, local citations, inbound links, and so on—also factor into how a page finds its rightful place on the Internet.

Plus, keywords are still relevant for your own purposes in thinking about common content themes and topics you want to cover, for taxonomy (organizing content and potentially navigation), and for things like AdWords campaigns. After all, when someone asks you what your content is all about, you can no doubt tell them quickly: somewhere in that answer lies a keyphrase or two. Plus, it can also be helpful to know if the topics you write about are also popular topics that many others write about: the more popular a topic, the more difficult it will be to rank for terms associated with it. In SEO parlance, these have historically been known as broad-head and long-tail keywords, and are an indicator of ranking difficulty for your content topics.

Dr. Pete Markiewicz notes that long-tail keywords—those that have lower search volume but are generally more specific—are more sustainable. "You're uniting very scattered audiences at low search cost to sell products and services that would be prohibitively expensive to do otherwise (e.g., Orphan Drugs). In other words, SEM with long tail is *more* sustainable than SEM with popular keywords."

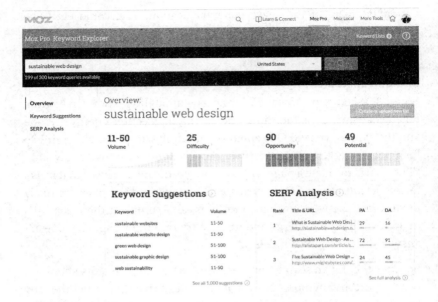

FIGURE 4-18.
Moz's Keyword Explorer can help you to discover and prioritize the best keywords to target

Keywords and phrases play a role when optimizing page tags, as well. However, for the keyword description metadata (where SEOs of yore used to stuff every keyword they could), not so much. That said, there are some practical places to optimize a page for a relevant keyphrase; however, make sure you do this after writing the best possible piece on the Internet for your topic (as discussed earlier in this chapter).

Here are some places where keywords and phrases still have relevance within a page:[10]

Meta description
> That snippet of copy that appears on search engine results page can make or break whether a user clicks through to your page. It should be clear, contain your ranking keyword, and be no longer than 155 characters.

---

10 Pratik Dholakiya, "Does Keyword Optimization Still Matter?", Convince & Convert. (*http://www.convinceandconvert.com/digital-marketing/does-keyword-optimization-still-matter*)

Title tag
> Keep it short and sweet (55 characters or less) and use your page's primary keyphrase.

Header tags
> Your H1 and H2 tags advise a search engine that the content within them is important and are great places to include keywords.

URL
> Include the keyphrase you plan to rank for at least once in the page's URL. Make sure it also describes what the page is actually about in a way that will be helpful to real users.

Content
> Yup, you probably included keywords and phrases in the body of your content without realizing it. They typically appear more frequently in content than other words or phrases.

Alt tags
> Including descriptive keywords in alt tags for your images not only helps search engines but also assists people with disabilities.

Finally, and perhaps most importantly, any place where keywords or phrases are read by users should sound *natural* and not obvious. Always make it human-friendly before robot-friendly.

Search on site

No conversation about sustainable search is complete without mentioning your site's own internal search engine. If you don't have one or if site visitors can't use it to find what they need quickly because it is not properly configured, you are not only doing your users a disservice, but also wasting energy. This might seem like a no-brainer, but you might be surprised at how much customization is required to get an out-of-the-box search plug-in for WordPress or Drupal module to behave intuitively for your unique content needs.

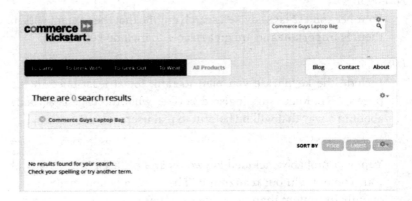

FIGURE 4-19.
Does your onsite search engine require a translator? Make sure the results it sends to users are relevant to their queries

Here are some questions to answer:

- Can the search field be used effectively on mobile devices?
- Does it properly interpret search queries that contain nonstandard characters?
- Do search queries include content that exists in third-party systems, like shopping carts, CRMs, content management systems plug-ins, or donation engines?

If your onsite search engine can't perform these tasks, you're frustrating users and wasting time and energy.

More sustainable social strategies

In 2014, we passed a tipping point of more mobile than desktop users on the Internet. As of August 2015, more than 2.2 billion of the planet's nearly 3.2 billion Internet users were active social media users with active mobile social users growing 23.3% in the previous year at an average of 12 new users every second.[11] Facebook alone has almost 1.5 billion users. It's clear we like sharing our cat videos.

But what does this mean in terms of sustainability? Well, according to Greenpeace's 2015 *Clicking Clean* report, not all social networks are created equal.[12] Whereas Facebook, Instagram, Snapchat, Pandora, Etsy, Flickr, and Google+, for example, are all either powered by data centers using primarily renewable energy or have made commitments to doing so, LinkedIn, Pinterest, Vine, Vimeo, Twitter, Tumblr, Reddit, Netflix, Hootsuite, Soundcloud, and many others as of this writing still have not yet transitioned, or even committed to, renewable energy. Many of these social platforms are hosted by Amazon Web Services (AWS), which we have covered in other chapters.

What is the trade-off between marketing and sustainability? Marketers want to drive people in droves to their sites, social media profiles, and other places where they publish. This, of course, drives up usage. Is this something content marketers and strategists should worry about?

Jen Boynton, Triple Pundit's editor-in-chief doesn't think so. "If a company were to ask my advice in this regard, I'd advise them not to worry about it as a part of planning," she says.

Her reasoning is that many complicated decisions go into creating content strategies and the data centers of social media companies are difficult to control or influence. "People's time and energy is very limited. Worrying about the emissions of a provider of a very small service like a Pin or a Vine is a distraction from larger sources of emissions that could be tamed."

---

11 Simon Kemp, "Global Digital Statshot: August 2015", We Are Social, August 3, 2015. (*http://wearesocial.net/blog/2015/08/global-statshot-august-2015*)

12 Greenpeace, *Clicking Clean: A Guide to Building the Green Internet*, May 2015. (*http://www.greenpeace.org/usa/wp-content/uploads/legacy/Global/usa/planet3/PDFs/2015ClickingClean.pdf*)

That said, she notes that there are other ways for content-driven companies to get involved and drive change. "If a company is concerned about this source of emissions, perhaps because social media is a huge amount of their business, we would advise them to advocate on behalf of energy efficient and renewable energy–powered data centers rather than avoiding using those problematic services," she says. "Advocacy can include direct action, such as publicly asking for change, and supporting the efforts of groups like Greenpeace to publish lists of bad actors and the like. This is a faster, more efficient route to change."

Orbit Media's Andy Crestodina encourages people to be as aware as possible of these issues, but take a holistic approach to how you address them. "There is a cost for our actions. When we don't see that cost, there is a high risk of big negative impact," he says. "Distributed costs with concentrated benefits: this is a formula for problems."

Andy recommends being very deliberate in all online actions. "Don't put something out there unless you believe in it and unless it matters to you, your marketing, and your audience. This applies to both marketers and consumers. Don't stream it unless you're consuming it. Don't share it unless you care about it."

FIGURE 4-20.
TweetFarts advice—every time you share, a gassy bird dies

Potential Barriers and Workarounds

All of the tactics in this chapter can help make your content workflows more efficient and use fewer resources in the long run, but there are still potential barriers when it comes to a more sustainable relationship with our content.

AWARENESS

Awareness is a big one. We have adopted a disposable mindset to all things Internet, *especially* when it comes to content. The general public remains blissfully unaware of the power and resources it takes to bring *Jessica Jones* or the latest J.K. Rowling tome to their iPad, even though in some cases the environmental impact could be worse than a printed counterpart or DVD. Content creators in the know should take Jen Boynton's advice and advocate for online platforms, Software-as-a-Service (SaaS) products, and social networks to embrace efficiency and, more importantly, power their servers with renewable energy.

Campaigns from Green America, Greenpeace, and others have already targeted companies like Amazon, Twitter, and Pinterest while also helping Apple, Facebook, Google, and others move more quickly toward renewable energy.

SHIFTING SANDS

Also, things are changing. For example, since 2014, Google penalizes nonadaptive/responsive websites in search engine results, giving mobile-friendly content a boost in the rankings. Plus, the semantic web and microformats continue to hold the promise of making search more intelligent and contextual. These advances will help users find what they need even quicker.

Related, in late 2015, Google worked with dozens of other publishers and technology companies to create the AMP project, an initiative to make web content load faster on mobile devices.

Similar to Facebook Instant Articles (closed source), AMP HTML (open source) offers a subset of HTML with limited and specific JavaScript features available. It is designed for content meant to be read rather than interacted with and offers functionality for including ad tracking and analytics code. "AMP is a fantastic industry-collaborative approach to make the mobile web faster," says Michael Ducker, Group Product Manager at Twitter who collaborated on the project.

In February 2016, Google integrated AMP Top Stories into all its mobile search results, giving mobile users faster access to popular search results. This is great news because only 5% of mobile media time is spent with the mobile web (the rest is in apps), according to data from Yahoo's Flurry unit. These low numbers are due to poor UX that results from slow page-load speeds, malware, and low-quality ads.

FIGURE 4-21.
AMP Top Stories started showing up in Google's mobile search results in February 2016

Conclusion

This chapter explored the basics of good search engine strategies and why those strategies help users find content faster. We also explored the importance of getting to the point quickly and telling an engaging story as well as content's role in helping users make more sustainable choices. In the spirit of good storytelling, we also touched on ways to make streaming video more compelling, efficient, and quicker to download. And finally, we talked about content marketing and the role sustainability can potentially play in sharing your story.

Action Items

Want to explore more of the ideas outlined in this chapter? Here are some things to try:

- Try to tell your story in 50 words or less. How about 20 words or less? In 140 characters?

- Create a content hypothesis and run a short experiment to check your hypothesis using the approach outlined earlier.

- Run your site or application through HubSpot's Marketing Grader or Open Site Explorer to see how it performs. Consider a plan for improvement. Begin by brainstorming some ideas.

- Want to know how difficult a particular keyword might be to rank for? SEMRush's Keyword Difficulty Tool can tell you in just a few clicks (*https://www.semrush.com/features/keyword-difficulty*).

[5]

Design and UX

What You Will Learn in This Chapter
In this chapter, we will cover:

- Good usability and its impact on sustainability.
- How best user experience (UX) design practices can help facilitate more sustainable design solutions.
- Just how many moving parts do you need in a more sustainable design process?

Users Versus Life Cycles
You're trying to book a ticket, make a donation, or simply subscribe to a newsletter on a device for which the content is not optimized. This is not only frustrating but wastes energy, as well. "An efficient website that is harder to use actually results in reduced sustainability, if you count the effort (and battery) life burned up by navigating through a lousy interface," says author and professor Pete Markiewicz.

In his book *Design Is the Problem*, author Nathan Shedroff suggests that designers embrace three strategies to create more sustainable designs:

Dematerialization
> This is the process of reducing the amount of materials and energy in a design solution; for example, simplifying web pages to use just the amount of design elements to achieve goals and no more.

Transmaterialization
> This is the process of transforming products to services, which can reduce the use of natural resources because services can be inherently less resource intensive. Think subscription-based online accounting software, for instance.

Informationalization
> This replaces the transportation of physical products (paper, for instance) with information, as in online billing, banking, email, and so on.

These three strategies conserve resources and reduce energy use, but don't specifically address efficiency, renewable energy, or the importance of helping users make more sustainable choices. Author Kem-Laurin Kramer notes in *User Experience in the Age of Sustainability* (Morgan Kaufmann, 2012), "it is important to design solutions that are responsive to the real needs of end users and at the same time strive to encourage more sustainable usage of the products."

Plus, as Pete Markiewicz points out in his article "Save the Planet Through Sustainable Web Design," not all dematerialization is good, because pixels are real and take up energy over time: "light as they are, those dots on the screen remain physical, and their survival requires a constant input of energy.[1] While some products become more sustainable by converting them to a swarm of bits, we must remember that those bits require something very physical to exist—a big, high-tech network, using lots of electricity and always-on computers to operate."

FIGURE 5-1.
Behind every good experience is a data center pulling electricity

---

1 Pete Markiewicz, "Save the Planet Through Sustainable Web Design", Creative Bloq, August 17, 2012. (*http://www.creativebloq.com/inspiration/save-planet-through-sustainable-web-design-8126147*)

In her book *UX for Lean Startups: Faster, Smarter User Experience Research and Design* (O'Reilly, 2013), Laura Klein states, "Lean UX isn't about adding features to a product. It's about figuring out which metrics drive a business, understanding what customer problems we can solve to improve those metrics, generating ideas for fixing those customer problems, and then validating whether or not we were correct." The same holds true when creating more sustainable design solutions, but we're looking beyond even business metrics or marketing goals to include the entire life cycle of the products and services we design. And that's a key difference.

Finally, as some sustainability practices rely on compliance and heavy documentation, which layer on regulations regardless of user needs, we need to ask ourselves, what is in line with and what is at odds with the methodologies of Lean/Agile UX?

BEYOND USERS: THE ENTIRE EXPERIENCE

Let's explore that broader view and how it applies to UX. As is discussed in Chapter 3, when considering sustainability as part of the UX design process, we must consider the entire life cycle of our products and services as well as the ramifications of our choices. Understandably, many UX practitioners focus exclusively on features and design elements that directly affect end users. Often, bigger-picture complexities are handled by product or program managers. But this is a stack with many layers. There are the UX and data layers, with which most practitioners are familiar:

Presentation layer
    This is how the experience looks and feels; its visual design language.

Task layer
    This is how the application flows and how you interact with it.

Infrastructure layer
    This is what the base technology product uses. Does it help or hinder UX?

And then there is the cultural and socioeconomic stack that makes it possible for us to do our work:

Device manufacturing
> Do the devices we design for use conflict minerals? Are they created with fair labor practices? How much hazardous waste does the manufacturing process produce?

Power sourcing
> Does the electricity that powers our applications come from renewable sources?

Device disposal
> Are we creating products for devices with built-in obsolescence? Does our work support or hinder that obsolescence?

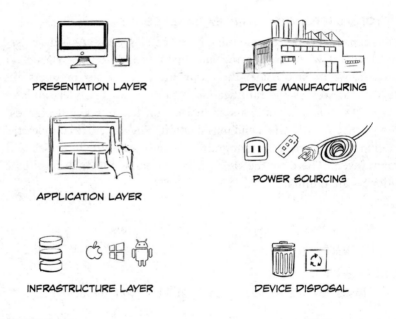

FIGURE 5-2.
Stacked up: considering the entire spectrum of UX

To truly integrate sustainable thinking into our process, the most important question to ask ourselves is this: Are we using these stacks to design solutions that promote accessibility, renewable energy, and break down socioeconomic barriers, or are we perpetuating a system of inequity? As a UX designer, it might be easy to say that conflict minerals and unfair labor practices, for instance, are not your problem (but most likely you wouldn't be reading this book if that were your stance). If you are a product manager, they are *precisely* your problem. As Green America's content strategist Bernard Yu pointed out in his presentation at the 2016 Sustainable UX conference: "Where in our UX process do we not only figure out how the products we're designing are made ethically, but even whether or not we are trying to solve the right problem?... Where do we step back and say it's not the product that's broken, it's the entire premise?"[2]

SUSTAINABLE DESIGN WORKFLOWS

With these big-picture questions in mind, how do we fit sustainability goals into design workflows to create more sustainable experiences across the board? Design teams should shepherd the process of helping both users and their clients make more sustainable choices across the entire life cycle of a product or service. But we should ourselves also create more sustainable solutions with efficient workflows and streamlined user flows.

UX designer, Amber Vasquez, advocates for dovetailing the two: giving users the best experience while also helping them make more sustainable choices.

It's about collaboration

Online creative portfolio community Format surveyed more than 2,000 photographers, designers, illustrators, artists, filmmakers, and other creative pros to understand how they prefer to spend their time.[3] In the survey, "49% prefer to share their work when it's fully complete, over sharing it when it's an idea or a work-in-progress." This is a problem.

---

2 See "Building More Holistic View of Sustainable: Digital Project Planning at Green America", available on YouTube. (*https://youtu.be/IW_qRxcQIe8*)

3 Format, "How to Live Like a Creative". (*http://format.com/creative-people?platform=hootsuite*)

If you run a search on the most common complaints about web designers, there is consistency in the results: "They disappeared." Or "I couldn't get in touch with them." Or "They didn't communicate enough." To make the design process both successful *and* efficient, you simply must collaborate. Often.

20th-century waterfall thinking won't work in the age of sustainable UX, where ongoing shared learning is valued and preparation trumps planning every time. Consensus from all parties must be reached at consistent touch points throughout a project's life cycle in order for it to stay on track. This is just as critical for software teams as it is in agency/client relationships.

FIGURE 5-3.
You can't have a successful project without ongoing collaboration

As Pete Markiewicz notes on his blog, "sustainability is best achieved by a team of 'hybrids' who know each other's work. This allows issues like efficiency to come up through the iterative design process. In a hybrid world, the hardcore coder may not be a great designer, but they have taken some design classes and understand why design is important. The designer may not be a hardcore coder (though hopefully they understand HTML, CSS, and JavaScript), but they understand it well enough to design for its effective use."[4]

AGILE PRACTICES, UX, AND SUSTAINABILITY

According to Philip Clarke and Zach Berke at Exygy, a software company and certified B Corp based in San Francisco:

> Agile is completely predicated on the idea of reducing waste: wasted time, wasted effort, wasted code, etc. If you are doing Agile effectively, you are always focused on the most high impact problem. As long as you are doing that well, you won't write code that won't get used. You will be in better communication with your team. You will get to a product that you will be able to ship sooner. It helps you get user feedback and start iterating faster than otherwise.

To repeat a common theme of this book: *Less waste = more sustainable.*

Streamlining user task flows for efficiency becomes one of several design imperatives when creating more sustainable digital products and services. If we consider the preceding strategies while auditing a website for sustainability, we are led to some of the following questions:

- Is the design process itself as efficient as it could be?
- Is a product or service's navigation and architecture as intuitive as possible to target users?
- Are design teams designing with real content? (see Chapter 4)
- Are labels and messaging clear and concise?
- Are all screens optimized for a specific conversion?
- Can users accomplish tasks in the least number of steps possible?
- Can users have a rewarding experience on the site despite device or platform?
- Is the solution solving a real-world problem?

---

[4] Sustainable Virtual Design, "The Green Team, Part I – The Role of Art Direction in Sustainability". (*https://sustainablevirtualdesign.wordpress.com/2012/03/09/the-green-team-part-i-the-role-of-art-direction-in-sustainability-39*)

PRACTICAL, TACTICAL

In the 20th-century, design took a top-down, one-way approach, but 21st-century sustainable design is a collaborative conversation. Here are some tactics to follow that can help make your design process more efficient and hence more sustainable.

Define your users, define their devices

When auditing digital products and services for sustainable usability, designers face an always expanding array of notebooks, laptops, tablets, phablets, phones, appliances, vehicles and other platforms on which their content must perform well for users.

Lean personas

You want to be sure the audience you're trying to reach is interested in what you plan to build, but with hundreds of tools at your disposal to extract behaviors and preferences from social media profiles and other online sources, it can be easy to fall into the rabbit hole of design research. Doing too much research wastes time and resources. Not doing enough risks missing the mark completely.

"Research is a big carbon cost in our projects due to amount of travel we do to where they are," says James Christie, director of experience design at MadPow. "I favor remote methods when possible. But the cost of not doing research and so wasting effort by requiring redesign is too high to risk not doing it."

Amber Vasquez notes that in general she doesn't think designers or agencies are doing enough research. When trying to pare down budgets, research and testing are often the first things to go. In many cases, user research isn't part of a budget, and that undermines project potential.

To compensate for this, she sometimes relies on "proto-personas," or lean personas, for projects where time or budget for research are restricted. Lean personas are based on hunches rather than hard research and serve several purposes: to jump-start the research process, to facilitate discussions with clients, and to align business goals with user needs.

The design team starts these lean personas based on their understanding of project requirements. The personas are then workshopped with the client, who fills in gaps based on its often deeper understanding of its users.

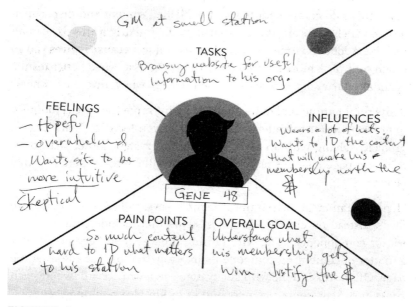

FIGURE 5-4.
Proto personas are helpful tools for quickly defining user needs

This process often occurs alongside a matrix exercise that compares client business and marketing goals with user goals to find the common ground. Finally, the design team fills in any remaining gaps with actual research from a variety of sources: online networks like LinkedIn or Facebook, user interviews, and so on.

The point is to run through the process quickly and efficiently through collaboration to glean key insights, and then fill in the gaps with more time-consuming research efforts, rather than the other way around. By starting with online research tools, it can be easy to get lost in the data.

Personas are meant to gain consensus on who you're designing for so that this understanding can inform design decisions throughout the process. Overthinking them wastes time and energy.

Lean wireframing

With many wireframing tools, you can get very granular in the amount of detail you include in wireframes. But at what point does all that detail become too much? How much time is wasted going back and forth about the placement of a button or slider, especially when it is likely that said component has an established content and display pattern to which users are already accustomed. Similarly, a one-size-fits-all tool like Adobe Illustrator might offer less value, because it gives you so much control. A more focused tool like InVision can yield better results faster. A quarter of a century into the Internet, why reinvent the wheel?

Some might argue that more detail in a wireframe leaves less room for interpretation (or misinterpretation), but if you focus on collaborative problem solving with an established design system rather than overthinking page element placement, you can likely reach better results faster.

"Rather than create wireframes," Amber says, "whenever possible we focus instead on creating design systems that include pattern libraries for components." The wireframing process then becomes simply a matter of combining components in a way that fits the intent of the page. Often this is done with sticky notes on a whiteboard (where each sticky note represents a component) or by simply sketching them out. Let's explore this a little deeper.

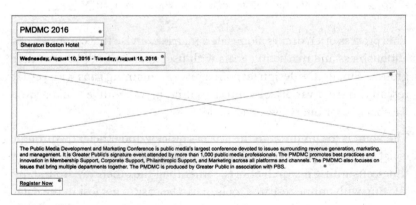

FIGURE 5-5.
A content pattern meant to display event info

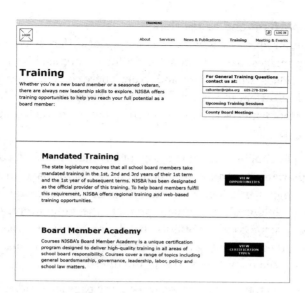

FIGURE 5-6.
A wireframe contains way more detail than a content pattern

Content patterns and page briefs

Every site, product, or service is made up of different types of content: hero images, banners, news feeds, forms, events, blog posts, text prompts, videos, white papers, and so on. These content types are often displayed in different ways, depending on the page upon which they exist. Content patterns show how the information for each content type is displayed.

For example, suppose that you are a company that throws a lot of events. You might want event announcements on your home page, in a sidebar on your blog, or in a footer on strategic pages. Even though all these announcements could potentially promote the same event, they will each have unique content patterns that define their size, type, and layout. Often, these patterns will seem instantly familiar because they are commonly used on websites all over the Internet.

Page briefs, on the other hand, do work at the page level rather than the individual content-type level. A page brief works exactly like it sounds. It's a short document that describes each page's purpose and includes specific conversion goals. Page briefs help both client and agency answer several important questions:

- Why does the page exist?
- What is the page's conversion goal?
- What is minimum amount of content that users need to see to make that conversion?
- What is the list of supporting information?

Page briefs help content creation teams achieve clarity about their own resources or constraints. In creating them, the content workload becomes much clearer. Page briefs also give design teams a helpful set of guidelines by which to create page layouts with designed components.

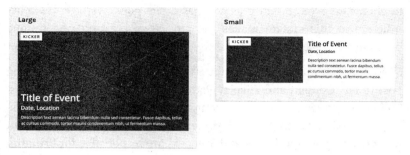

FIGURE 5-7.
Because they are based on common interactions across the Internet, content patterns are familiar and easy to use

Display patterns and component design

A content pattern is different from a display pattern, which shows how that content type will be displayed to fit within the visual design system defined during the style tile process. Components are concrete implementation of design patterns. With display patterns and component design, rigid full page templates are less necessary. By building a library of content and display patterns that can be applied to specific components, you can create a modular design system that lets you mix and match components for optimal flexibility.

FIGURE 5-8.
The component library Mightybytes created for NACSA let it mix and match components on its own to fit unique content needs

After you have a library of approved patterns and components, directives from your page briefs will help you choose which are most relevant to suit the page's conversion goals. Component design is beneficial to clients, as well, because as soon as a site goes live, the flexibility inherent to this system makes the process of creating new unique pages much easier.

So how do we make this work for a team? Who does what? In a blog post titled "Content and Display Patterns," author Daniel Mall notes, "When thinking about patterns, content strategists are primarily thinking about Content patterns; designers are primarily thinking about Display patterns; and frontend developers are responsible for bringing the two together."

User stories

For Agile teams, user stories are critical for defining actual features and functionality required for a product roadmap. They typically take this form:

As a <role>, I want <feature>, so that <reason>.

An example of this might be,

As an *admin*, I want *to add new users*, so that *they can update the blog*.

During discovery workshops, product teams collaborate to define the necessary features and functions that define the product. After these features are defined, they are prioritized and categorized. Design and UX teams then use these stories to define content patterns, wireframes, and components that make up the product or service's design system.

Critics of user stories have said that they are not detailed enough to truly describe a feature or function, that they can be ambiguous or leave room for interpretation, especially in the design process, and that can lead to scope creep.

As part of a larger Agile framework, user stories can make your process more efficient, which saves time, resources, and energy. Some teams like the tactile flexibility of using sticky notes and whiteboards, whereas others prefer an online sorting tool like Trello. If you're using digital tools, they of course require electricity to run and Trello, as an example, is hosted by Amazon Web Services (AWS), which doesn't get high marks for renewable energy use. If you're going analog, consider recyclability of your materials. Even though many sticky notes are made of recycled materials, the adhesive backing can cause them to be rejected by some recycling programs. If you use a lot of stickies, go with a recycling program that supports mixed paper. Also, in America alone, 400 million whiteboard markers are disposed of each year. AusPen markers use nontoxic inks and are refillable, making them a more environmentally friendly alternative.

Avoid dark patterns

Run into the light, Carol Anne! The website Dark Patterns is dedicated entirely to pointing out UX patterns that deceive users.[5] These deceptions waste time and energy and cause untold frustrations. They're just bad news all around. Beyond basic design process no-nos, which are just annoying to users, like asking for things you don't need or letting an IT or HR department design your forms, dark patterns are *intentionally* deceitful.

Here are some things to avoid:

- Don't ask for credit card numbers for free trials and then make it difficult to unsubscribe to a product or service.

- Similarly, don't make it easy for users to subscribe but then difficult to unsubscribe. (This is called a *Roach Motel* in UX parlance.)

- Don't misdirect users by changing patterns without warning. In other words, if the blue button consistently takes you to the next screen but then suddenly becomes a "Buy" button, users will instinctively click it before realizing that it does something different than the other 19 buttons of the same size and color. Users don't want to feel they've been tricked into taking an action, especially not if it requires spending money.

- Don't disguise advertisements as other kinds of potentially more desirable content. 'Nuff said.

- Don't add surprise or hidden costs at the last step in your checkout process. Similarly, don't slip unwanted items into a user's cart.

- Don't take advantage of the fact that users scan copy on the Web rather than reading in detail, as in the case of disguising opt-out buttons as opt-in buttons with misleading copy.

---

5 *http://darkpatterns.org/*

Give it a Rest

In the article "Design through the Twelve Principles of Green Engineering," the authors P.T. Anastas and J.B. Zimmerman write that "Products, processes and systems should be 'output pulled' rather than 'input pushed' through the use of energy and materials."[6] In other words, don't make your product run when it doesn't need to. Give users control over functions whenever possible. If a product or service has built-in notifications or backups, for example, let users control how often that happens and keep energy use in mind when making default settings. This is a great opportunity to highlight more sustainable choices in your designs as well. Your product will use far less energy checking twice a day for new data than it will every five minutes.

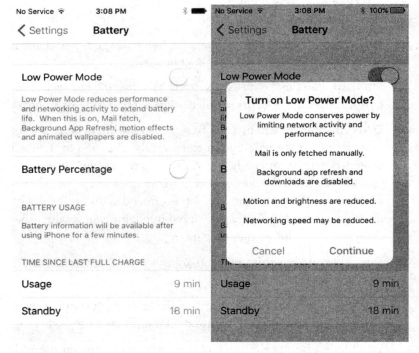

FIGURE 5-9.

Let users decide when their product goes to sleep or how often it sends notifications

---

6 Paul Anastas and Julie Zimmerman, "12 Principles of Green Engineering". American Chemical Society. (https://www.acs.org/content/acs/en/greenchemistry/what-is-green-chemistry/principles/12-principles-of-green-engineering.html)

If an app uses a phone's camera, WiFi/mobile data, and GPS capabilities at the same time, as in the case of Snapchat, it can significantly impede performance, drain batteries, use high amounts of data traffic, and waste a lot of energy. Keeping this in mind during UX exercises will also encourage conversations about usability and energy efficiency to happen in tandem with each other.

Visual Design

Visual design choices inform sustainable UX, as well. Here are some ways to streamline visual design workflows and the deliverables that come from them.

YE OLDE DESIGN COMP

Like many design firms, for years Mightybytes delivered standard design comps—pixel-perfect visual renditions of key site pages—for website projects. Getting client approval on those comps was a critical part of moving projects forward. During this process, we sometimes became stuck in extended design revision loops, or when we moved to the build phase, complications around the approved design comps arose. Occasionally both happened. Even when we added wireframes to the mix, we still ran into challenges. Clients were often understandably confused by the process of reviewing just a static screen or two and projecting what the end result might entail. Design comps only represent a single state of a single page of your website or application, so this is not surprising. They're a snapshot in time and do little to help clients or users understand the broader implications of choices made long before there is anything with which to interact.

"It's easy for people on both client and agency side to fall into the Stockholm Syndrome trap of waterfall-based design," says Amber Vasquez. "It's human nature to apply old methods to new mediums, but the reality is that rarely works. People still try to shoehorn the print design process into web projects, but time and time again they fail, or at the very least go wildly over budget."

Still, imagining a project without design comps can be hard for clients to wrap their heads around. Our presale process for helping them better understand a more efficient way of designing digital products and services is as follows:

- We give a brief overview of the process in early conversations.
- We include a step-by-step breakdown with example deliverables of our component-based design process with every proposal.
- We link to blog posts that we (and others) have written that show clear benefits and mutual value for our team and end clients.
- When applicable, we include testimonials from clients who have personally reaped the benefits of this process.

Let's look at this way of doing things in more detail.

STYLE TILES

At Mightybytes, some of our earliest design deliverables are *style tiles*, which communicate the essence of an online visual brand. More detailed than mood boards and designed specifically for websites and applications, style tiles include the most common visual assets of a website: font, colors, image styling, interface elements, and so on. Their purpose is to help client and agency find a common visual language with which to design the various components of a digital product or service. Because they are independent of layout or UX, style tiles can help agency and client teams gain consensus around how site components will be styled. Style tiles form the basis for creating layouts.

If the client has an existing style guide or brand book, that can be very helpful for creating style tiles. Font choices and color swatches already exist for other mediums. Design teams create style tiles to translate how those guidelines will work on the Web. They also create styles for iconography, image treatments, buttons, and other common interface elements. Those elements then form the basis of styles defined in cascading stylesheets, which control the look and feel of an entire site. If it is decided that header fonts should be gray instead of black, that's a thing that can happen quickly.

FIGURE 5-10.
These style tiles helped the team at Bike & Build better understand color schemes, image stylings, and button treatments well before they were applied to site components

COLOR CHOICES

In the early 2000s, sustainability-minded designers suggested using darker colors in digital design to reduce the energy required to power white or brightly colored pixels. In fact, there was even a Google search spinoff created by Heap Media, called Blackle that, as of this writing, purports to have saved 5,461,326.853 Watt hours of electricity by allowing users to search Google on a dark rather than white screen. Whereas this energy-saving effort held true when most monitors still used cathode ray tubes to beam red, green, and blue phosphors across your screen, LCD monitors or LED-lit LCD screens can actually use more energy to display a black screen than a white one.[7]

---

7 Larry Greenemeier, "Fact or Fiction?: Black Is Better Than White for Energy-Efficient Screens", *Scientific American*, September 27, 2007. (*http://www.scientificamerican.com/article/fact-or-fiction-black-is*)

But LCDs aren't the only screens out there. In OLED screens, found in newer smartphones and tablets, each pixel produces its own light (rather than using a backlight like LCD monitors), so if the primary color of your design is black, no pixels are activated, which results in much less energy use.[8] However, white-on-black screens are often harder to read, introducing an accessibility problem. So although it might be lower energy with OLED, if people have to peer at a page for twice as long to make sense of the content, it's a net loss.

Given this, when designing interface screens does it matter which colors you choose? The answer depends on two things:

- How well you know your audience.
- The importance of brand consistency in your design work.

First, hundreds of products, including many tablets, devices, and smartphones by Samsung, Panasonic, and others use OLED screens.[9] As of this writing, the iPhone still uses an LCD screen.[10] Design choices that are energy efficient on one type of screen use more energy on another type of screen. If you know that the majority of your users will be on a specific type of device, you can make an educated choice, but how often does that happen?

Similarly, accessibility guidelines for people with low vision might require making a high-contrast stylesheet. A style tile might have a set of reduced, high-contrast colors for this use case.

Second, colors are important enough to any organization's brand that it is very likely brand consistency will trump energy efficiency in most scenarios. A potentially more useful decision might be to make use of "flat" assets—images that contain large swaths of flat color rather than gradients, bevels, or other effects—in your design work. File

---

8 Quora, "Does a White Background Use More Energy on a LCD Than If It Was Set to Black?". (*https://www.quora.com/Does-a-white-background-use-more-energy-on-a-LCD-than-if-it-was-set-to-black*)

9 OLED-Info, "OLED Products: Comprehensive Guide". (*http://www.oled-info.com/devices*)

10 Dr. Raymond Soneira, "The iPhone 6 and 6 Plus Have the Best LCD Screens You Can Buy", Gizmodo, September 22, 2014. (*http://gizmodo.com/the-iphone-6-and-6-plus-have-the-best-lcd-screens-you-c-1637612720*)

sizes for CSS buttons and vector-based SVG files are smaller than their raster-based counterparts, meaning less data to download. Plus, flat images can be scaled more easily in a responsive design.

FIGURE 5-11.
Which color choices uses the most energy? It depends

FONTS AND TYPOGRAPHY

Designers have thousands of typographical choices to use in their digital products and services. How sustainable they are depends on two things:

- How much load time they add to your page.
- How many HTTP requests they add to your page.

Let's take a look at common font types and the pros and cons of each:

System fonts
> These fonts exist by default on the user's machine/device and use the least amount of resources, making them automatically a more sustainable choice. They also restrict design choices because most systems only have a few such fonts installed.

Web fonts
> Web fonts offer nearly unlimited choices and can be displayed on devices where the font isn't actually installed. Because they are hosted on external servers, they increase the number of HTTP

requests your page makes and can thus affect its load speed. In experiments we ran at Mightybytes, adding a font through Typekit added an average of 11 KB to a page, whereas Google Fonts added about 28 KB.[11]

Embedded fonts

Using these types of fonts, you can embed fonts into your page using the CSS style @font-face. Although embedding a font increases page weight more than the other two options, you can use a tool like Font Squirrel to optimize the font by removing elements you may not use like dingbats or specific foreign language characters. You can also use Base64 encoding to convert the font to a data URI that can be embedded directly into your stylesheet. Even though it will make your stylesheet larger, it will reduce HTTP requests.

WEB FONT   **SYSTEM FONT**   **EMBEDDED FONT**

FIGURE 5-12.
System fonts, web fonts, embedded fonts: which best suit your needs?

Here are some other good practices to consider:

- Keep it to two fonts maximum—there is no need for many web fonts.

- Web fonts are often used for icons (e.g., Font Awesome), but this approach is considered flawed now—use SVG for icons, not icon fonts.

- Consider typefaces designed for screen, rather than print; for example, Georgia or Verdana.

---

11 Amber Vasquez, "Web Fonts and Sustainability", Mightybytes Blog, July 19, 2013. (*http://www.mightybytes.com/blog/sustainability-web-fonts*)

- Consider making the majority of your design system fonts and using webfonts sparingly. No using three kinds of web fonts. Find the right mix.
- Reduce the number of typeface weights in a given kit. You most likely don't need them all.
- Finally, hosting a font on a green host is better than being served by Adobe's Typekit. At least you have control over download.

Weigh your options carefully when making typographical choices for your digital product or service. If the project calls for high design, go with a web font. If it requires optimal performance, perhaps a simple system font will work just as well. Or, maybe the web font works well for desktop users but a system font works on mobile devices. Just make sure that the font options you choose don't impede performance. If you're not sure which choice to make, refer to your page weight budget. Do you have the extra kilobytes to spare?

IMAGERY

Let's explore some things you can do with visual assets on your site, such as photos and icons, to keep file size down while still maintaining visual integrity.

Content size by content type

CONTENT TYPE	PERCENT	SIZE
Other	56.7 %	3.86 MB
Script	19.2 %	1.31 MB
Image	18.7 %	1.27 MB
{ } CSS	2.4 %	170.47 KB
Plain text	2.1 %	149.27 KB
HTML	0.8 %	55.22 KB
Total	100.00 %	6.81 MB

Requests by content type

CONTENT TYPE	PERCENT	REQUESTS
Image	37.5 %	155
Other	28.1 %	116
Script	25.9 %	107
HTML	2.9 %	12
{ } CSS	2.9 %	12
Plain text	2.7 %	11
Total	100.00 %	413

FIGURE 5-13.
Images add both visual impact and weight to your page—use them wisely

Images can add a large amount of extra data to your page. First of all, ask yourself if you even need an image. Is a photo or other sort of image critical to telling your story or teaching a concept inherent to the success of your content? If you decide an image is absolutely necessary, can you achieve the same result using resolution-independent CSS effects or even web fonts?

Which format?

Vector image formats such as SVG are resolution-independent and can scale up or down without adding additional weight to your page. These are great for shapes, icons, logos, and so on. But they are completely unsuitable for complex imagery like photos. That's where raster images like PNG or JPGs are better. Raster images use pixels on a grid to display an image. Because of the way they are constructed, raster images will also be bigger files than their vector counterparts, so it makes sense to consider the display need carefully as you ready files of web delivery. Here are some guidelines to keep in mind.

- PNG or GIF files are great for diagrams or graphs that have flat line art and don't need to scale. PNG files also support alpha transparency.

- JPG is a great option for photos, but logos or other line art will look jaggy because of JPG's compression scheme.

- SVG, as just mentioned, is great for resolution-independent line art with flat colors (gradients, not so much). When scaled up, they will retain their sharpness. Because vector images are stored as geometric equations, they can also take more computing power to render. They also offer more flexibility because they can be animated or controlled with CSS for richer interactions.

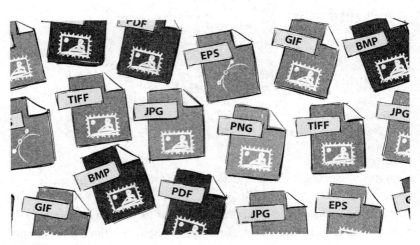

FIGURE 5-14.
Which image format is best for your needs?

Compress those images

Be sure to run all of your images through a compressor like Smush.it, Kraken.io, or Image Optimizer to shave off extra file size. Choosing "Save for Web" in an image editing tool like Photoshop can add extra unnecessary metadata and export files up to 50% larger in size than other optimization tools.

Also, from a visual design perspective, black and white photos or illustrations are both less data-heavy than retina-ready JPGs.

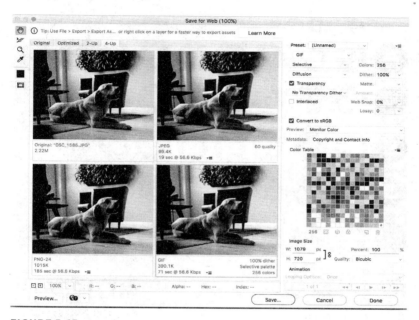

FIGURE 5-15.

Compression matters: strive to find that perfect balance between quality and file size

Use CSS sprites

CSS sprites combine multiple graphics or images into a single file and use CSS properties to display only the portions necessary for a specific need. With this technique all your site's icons, for instance, can be downloaded once, cached by the browser, and displayed across all screens without incurring additional HTTP requests or the additional data size associated with multiple small images.

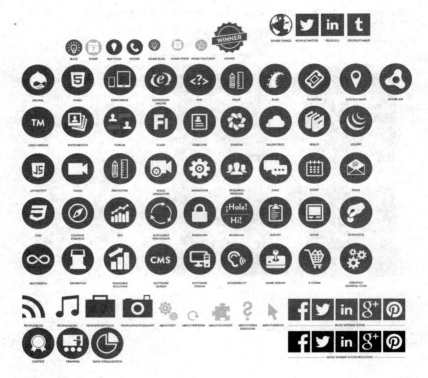

FIGURE 5-16.
One image, many uses. CSS sprites cut down on HTTP requests and file size

Inline images

In some cases, it might be helpful to inline an image right into your HTML. Using Base64 encoding (which, having been first proposed in 1987, predates even the World Wide Web), images can be included directly in your HTML, reducing HTTP requests and allowing pages to load faster.

Several Base64 image encoders are available for free online. Simply drag and drop an image to convert your image into a Base64 file. Some will also compress images before reencoding them as a Base64 file.

This approach is best used on small, single use images. Encoding large photos, even when using compression (which you always should), will result in slower page loads because Base64 encoding doesn't reduce the actual amount of data being sent to the browser. It only reduces the lag time caused by HTTP requests. This approach is also better for images that are used only once rather than those commonly reused, like icons, or navigation elements, because they don't cache in the same way a CSS sprite would.

Also, if images are an important part of your SEO strategy, this is not the way to go, because Base64-encoded images will not show up in search engine results at all.

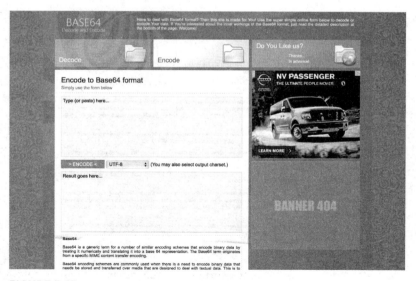

FIGURE 5-17.
Patrick Sexton's Base64 image encoder will turn your image into a data stream

PRINT STYLES

Yes, people do still print web pages—not often, but it happens. Recipes, bank statements, contracts, FAQs, maps, directions, and so on are all prime candidates for printing. How do your screens look when printed? One of the quickest ways to waste massive amounts of paper, ink, and other resources is to print pages that aren't optimized for a printer. Sustainable design principles dictate that printer-friendly screens or web pages use the least amount of resources while still providing acceptable results.

To do so, print should be treated as an equal alongside other media types, like screen and speech. Although most browsers will by default change colors automatically to save ink, this is often not as effective as a custom solution. You can use media queries in CSS3 to look at the capabilities of a device and check a number of things, including width/height of viewport or device, landscape or portrait orientation, or screen resolution. Using these queries, you can call individual style rules or separate stylesheets to adjust your page based on different criteria.

FIGURE 5-18.
You might not realize it, but people do print your pages; not optimizing them can lead to wasted paper, ink, time, and other resources

Print stylesheets, provided as additional CSS files for a site, don't need to include all relevant styles, only the differences between screen and print. If you use the print media query, default styles will generally be included. It is important to define the differences clearly. For example, while pixels are the default unit of measurement for screens, for print it's inches or centimeters, so include things like margins and padding in those units.

Here are some things to consider when creating print styles:

Backgrounds, images, and colors
  Use print styles to force the color of a specific page element to be more printer-friendly and use less ink. For situations in which an image or background is critical, you can also override default browser settings to force render them. Similarly, you can use CSS filters to reverse images that include light text over dark backgrounds.

Navigation and buttons
  Because these are not relevant to a printed page, remove them by setting the display in the print style to none.

Images
  You don't want images to bleed over the edge, especially if they force another page to print with just a small section of the image on it. Setting the maximum width to 100% on images will keep this from happening.

Expand outbound links
  Make sure the link URL is shown for outbound hypertext links.

Include a print button
  Using a print button that specifically calls upon the print stylesheet will produce better results than using the browser's default print settings.

The important thing to remember is to keep the file size small while still maintaining acceptable print quality. Web teams often overlook print styles, but in doing so they unwittingly contribute to waste of resources. Taking a bit of extra time to make your pages printer-friendly can make a difference and reduce user frustration.

MEASURE SUCCESS

Finally, as we discussed in other places throughout the book, ground all your design efforts in a solid strategy for measuring success. Set goals early and work through strategies to decipher whether you are meeting those goals quickly. Make no assumptions on your design decisions or those requested by clients. Instead, build user testing into every step of the process if you can. By testing design decisions with real users, you can move your process forward based on hard data. This will help keep those uncomfortable conversations driven solely by subjective emotion—"But I *feel* it should be green"—at bay.

A/B testing—whereby two UX options are presented to users and interactions are tracked—can go a long way toward getting better design decisions based on data made faster.

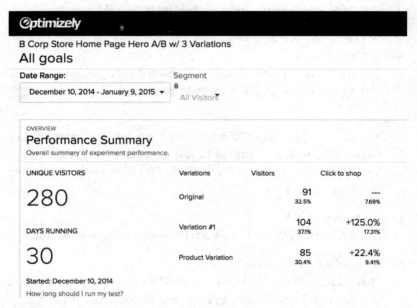

FIGURE 5-19.
Measure what matters: the most successful design projects are those that use data to support decision making

Resources like Optimizely, UsabilityHub, Optimal Workshop, and others can help move design decisions forward quickly without exorbitant expenditures on custom user testing workshops. Those of course have their place, but not all projects have the luxury of bringing a room full of target users together for an entire day of testing.

ACCESSIBILITY, SUSTAINABILITY, AND DESIGN

Ken-Laurin Kramer notes in *User Experience in the Age of Sustainability* (Morgan Kaufmann, 2012) that "Sustainable design should strive to be inclusive and 'universal' in its appeal. Through using universal design practices and following those design guidelines, we can practice sustainable design and accessibility in one effort." In other words, accessibility-compliant products and services that work across the widest array of computers and devices, including screen readers and other enabling technologies, are potentially more sustainable solutions than those that don't consider the *entire* spectrum of users.

When someone talks about a digital product or service being accessible, they're often referring to whether it works for people with disabilities. The World Wide Web Consortium (W3C) publishes the Web Content Accessibility Guidelines (WCAG), which can help product teams create more accessible and hence sustainable experiences.

WCAG includes twelve guidelines organized into four categories:[12]

Operable
> Users must be able to operate all interface components and navigation. Guidelines include:
> - Make all functionality available from a keyboard
> - Provide users enough time to read and use content
> - Do not design content in a way that is known to cause seizures
> - Provide ways to help users navigate, find content, and determine where they are

Understandable
> Content or operation of interface information cannot be beyond a user's understanding. Guidelines include:
> - Make text content readable and understandable
> - Make web pages appear and operate in predictable ways
> - Help users avoid and correct mistakes

---

[12] W3C, "Web Content Accessibility Guidelines (WCAG) 2.0". (*https://www.w3.org/TR/2008/REC-WCAG20-20081211/#contents*)

Perceivable
> Information being presented can't be invisible to all user senses. Guidelines include:
>   - Provide text alternatives for any nontext content so that it can be changed into other forms people need, such as large print, braille, speech, symbols, or simpler language
>   - Provide alternatives for time-based media
>   - Create content that can be presented in different ways (e.g., simpler layout) without losing information or structure
>   - Make it easier for users to see and hear content (e.g., separating foreground from background)

Robust
> Users must be able to access content as technologies advance and user agents evolve. Guidelines include:
>   - Maximize compatibility with current and future user agents, including assistive technologies

FIGURE 5-20.
Sustainable design should strive to be inclusive and universal

Web Standards

Web standards and sustainability go hand in hand just as sustainability and accessibility do. Using the standards outlined by the W3C, you can "build rich interactive experiences, powered by vast data stores, that are available on any device." Designers powering their work with web standards–friendly markup like HTML and CSS means they're supporting the widest array of devices, which makes their products and services more accessible and, if their markup is also stripped down to the bare essentials without compromising UX, more sustainable.

The W3C breaks its full list of standards down into these categories:[13]

Web design and applications
> These are the standards for building and rendering web pages. Includes HTML, CSS, SVG, Ajax, and so on.

Web of devices
> These are the technologies that enable Web access anywhere, any time, using any device. Devices can range from smartphones to wearable technologies, interactive television, automobiles, and so on.

Web architecture
> These is the foundation technologies of the Web, such as URIs, HTTP, and so on.

Semantic web
> The technologies that enable people to create data stores, build vocabularies, and write rules for handling data using such technologies as RDF, SPARQL, OWL, and SKOS.

XML
> This defines standards relating to XML technologies, such as XML, XML Namespaces, XML Schema, XSLT, EXI, and others.

Web of services
> Message-based design services based on technologies like HTTP, SOAP, SPARQL, WSDL, XML, and so on. Frequently found in technologies related to payments, security, and internationalization.

---

[13] W3C, "Standards". (*https://www.w3.org/standards*)

Browsers and authoring tools
> The tools we use to access and create web content: browsers, media players, content management systems, social media, photo and video sharing apps, blogging tools, and many others.

Though a deep dive into web standards is beyond the scope of this book, for web designers and UX practitioners, making sure your digital products and services are designed with standard HTML and CSS markup will ensure that the widest array of users can access your content. This includes search engines (see Chapter 4), so optimized content wrapped in standards-based design is also more easily findable. Standards-based pages will also perform better, serving up information more quickly to users. Because web standards are written to be both future-proof and make your work backward-compatible, they will be more stable and reliable, as well. These are all things that lead to higher levels of sustainability. Score!

Now let's explore some other ways in which design tactics can yield more accessible and hence more sustainable results.

MOBILE-FIRST

Mobile-first is a term coined in 2009 by Luke Wroblewski to define a design strategy for mobile devices.[14] It has three components:

- Mobile is growing exponentially and offers many new opportunities.
- Mobile requires design teams to focus on what's most important.
- Mobile offers new capabilities and opportunities for innovation.

These parameters require you to focus on content that's most important for your users. Mobile-first strategy requires designers and content creators to consider the needs of users on mobile devices before considering those on desktop machines, which typically have larger screens and more processing power. This follows the basic principles of progressive enhancement (which we'll cover next), the thought being that if something offers a great experience on smaller devices with fewer resources, that experience can be enhanced for users on more powerful devices rather than the other way around. Mobile-first forces a product

---

14 Luke Wroblewski, "Mobile First", LukeW Ideation + Design, November 3, 2009. (*http://www.lukew.com/ff/entry.asp?933*)

owner to consider content and interactions that are most essential to meeting its goals and user needs before adding any additional bells and whistles. This can be a helpful exercise for clients who cling to their content regardless of how it performs.

When designing a digital product or service using mobile-first strategy, content patterns and component design become critical steps to achieving success. With only so much real estate in which to display your content and interactions, agency and client need to work together to reach consensus on the most essential elements for each screen. This in turn informs design decisions for larger screens.

FIGURE 5-21.
Focusing first on what mobile users need allows for more scalability

Mobile-first is a more sustainable strategy because when content, transactions, and interactions are prioritized with mobile devices in mind first, end-user devices load less data, which means they consume less energy. Similarly, when visual design decisions are made with mobile-first in mind, mobile pages load more quickly while using less bandwidth, resulting in energy savings, as well.

PROGRESSIVE ENHANCEMENT

Progressive enhancement is a web strategy that applies technologies in a layered approach which gives everyone access to basic content and functionality and increases functionality as a user agent, such as a browser, will allow. In other words, the more modern and feature-rich a user agent is, the more features are revealed by a page that uses progressive enhancement.

This is relevant for designers and UX practitioners because rather than designing for graceful degradation—that is, creating experiences with the most bells and whistles for top-of-the-line machines first and worrying about smaller/older/slower devices as an afterthought—progressive enhancement flips the equation with a focus on maximizing accessibility. When designing with progressive enhancement in mind, consider the lowest common denominator first and create the best possible experience for that, and then layer additional enhancements atop that experience.

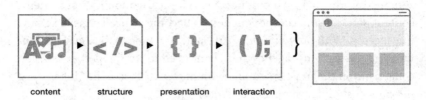

FIGURE 5-22.
Progressive enhancement insures that users on lower-end devices still have access to basic content and functionality

Mobile-first is a natural extension of progressive enhancement. Like the other strategies and approaches in this section, its focus on accessibility makes it also a more sustainable strategy.

RESPONSIVE DESIGN

The term *responsive web design* was coined by Ethan Marcotte in a 2010 article for A List Apart.[15] Responsive designs accommodate different screen resolutions using fluid grids that respond to a device's screen size by resizing accordingly. This is typically done by using media queries in CSS3, which give pages the ability to tailor themselves to a specific range of output devices without changing content.

Critics of responsive web design note that responsive sites can take longer to load because the media queries used to make a site responsive simply scale or resize assets based on the requesting site's viewport size. Thus, a responsive page on a mobile device could in fact be loading the same assets as it does on a desktop site, which is an unfortunate waste of resources and bandwidth. But when used in tandem with mobile-first and progressive enhancement strategies, responsive web solutions can deliver optimized experiences for users across devices, making them more accessible to more people while still using fewer resources.

---

15 Ethan Marcotte, "Responsive Web Design", A List Apart, May 25, 2010. (*http://alistapart.com/article/responsive-web-design*)

Similarly (but not quite the same), adaptive designs adapt to the width of the browser at specific points. In other words, an adaptive site is only concerned about the browser being a specific width, at which point it adapts page layout.[16]

An analytics tool will provide you with information about which pages have high amounts of traffic from mobile devices. If budget or timing are concerns (and when aren't they?) you can begin by tweaking those pages first.

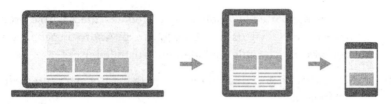

FIGURE 5-23.
Sometimes responsive designs just scale down desktop-sized images using HTML, which can cause lag times on mobile devices

Regardless of how you address the needs of mobile customers, doing so is absolutely critical. We have passed the tipping point: more content is accessed via mobile devices than not. More searches are done on mobile devices than on desktop machines. Thus, it's a safe bet that the majority of your users are on mobile devices. Plus, in April 2015, Google announced that its updated search algorithms would rank mobile-friendly sites higher than their nonmobile counterparts in search results. So you actually are penalized in search results for not having a mobile-friendly site.

---

16 Geoff Graham, "The Difference Between Responsive and Adaptive Design", CSS-Tricks, November 11, 2015. (*https://css-tricks.com/the-difference-between-responsive-and-adaptive-design*)

Potential Barriers to Sustainable UX

What are the things that can get in our way of designing more sustainable digital products and services? Designing a checkout system with too many steps is something you can fix, but what are the larger barriers between us and more sustainable design solutions?

Proprietary technologies are a big consideration:

- Flash and other proprietary technologies requiring plug-ins are not supported on many mobile devices.

- Java is not supported by Apple mobile products and Android products use a custom virtual machine to run an Android-specific version of Java, which could lead to many frustrated users.

- Any other technology that requires a plug-in or doesn't run under web standards.

Flash, for example, is still used by many online sites as a format for video playback, despite standards-based video players having been around for years now. It is also used by many advertising networks and can significantly slow down news sites and other popular ad-driven content sites. Plus, in addition to being a bandwidth and energy hog, as I just mentioned, most mobile devices cannot play Flash content, so if you want to create more sustainable user experiences, ditch the Flash.

Conclusion

We covered a lot of ground in this chapter:

- We encouraged UX designers to consider long-term needs and the entire life cycle of the products and services they build rather than simply short-term user needs.

- We talked about the impact good usability has on sustainability.

- We covered some practical tactics you can use to optimize design solutions for more sustainable delivery.

- We explored how accessibility and web standards affect sustainable design solutions.

In Chapter 6, we discuss how you can apply performance optimization techniques to your design solutions for faster, more energy efficient delivery of products and services.

Action Items

Try these things:

- Write-up some proto-personas about your users based on hunches. Do some research to decipher how close you were.
- Flesh out some content and display patterns on sticky notes. Use them to wireframe several pages.
- Use WebAIM's web accessibility evaluation tool, WAVE, to see how accessible your designs are to people with disabilities. (*http://wave.webaim.org*)
- Use Google's Mobile-Friendly test to check how mobile-friendly your work is.
- Use MobileTest.me to see how your site looks on a specific device (*http://mobiletest.me*).

[6]

Performance Optimization

What You Will Learn in This Chapter
In this chapter, we will cover:

- Why optimizing performance is a critical web sustainability component.
- Techniques for creating better optimized digital products and services.
- Workflow tips for assessing performance.

Performance Counts
Can digital products and services find the perfect balance between speed, reliability, and sustainability? In an interview, Chicago-based developer Eric Mikkelsen brought up this point about the dichotomy between them: "Speed and reliability are interesting because sometimes those things are opposed to each other," he says. "Building a website that takes advantage of techniques for tomorrow yet requires a fallback for yesterday increases the size of our codebase." So where does the perfect balance lie?

Several years ago, Amazon calculated that a page load slowdown of just one second could potentially cost the online retailer up to $1.6 billion in annual sales.[1] Similarly, Google noted that slowing search results by a mere four-tenths of a second would cut down search volume by eight million queries per day, depriving them of the opportunity to serve as many or more ads and resulting in a direct loss of potential revenue.

---

[1] Kit Eaton, "How One Second Could Cost Amazon $1.6 Billion In Sales", *Fast Company*, March 15, 2012. (*http://www.fastcompany.com/1825005/how-one-second-could-cost-amazon-16-billion-sales*)

There are many examples like these, and losses don't just take the form of customer revenue either. Netflix, for example, saw a 43% decrease in its bandwidth bill after enabling GZip, a common file compression tool.[2] It is clear that better performance yields better financial results and happier users.

In practice, maintaining a commitment to optimized performance can, apparently, be challenging. The average web page, according to the HTTP Archive, crested at more than 2.4 MB in May 2016. Bloated, rotating image carousels, slow share widgets, and looping video backgrounds have become the norm on websites everywhere, even though there is mounting evidence that these features can adversely affect page performance. Clearly, there's a disconnect somewhere. How does this happen?

In an article he penned for A List Apart titled "Sustainable Web Design," James Christie notes that "No one intentionally sets out to build a 1.4 MB page, but clients often demand eye-catching images, advanced social features, and plenty of design bells and whistles, and that's pretty much all it takes to get to that size." (At the time Christie wrote that article in late 2013, the average page size was only 1.4 MB, so you can see how quickly they are rising.)

It is easy to blame clients for bloated websites that perform poorly—they want what they want, after all, and are typically willing to pay for it—but building high-performance products and services is a cause we should all get behind, including clients. Keeping our clients informed and guiding them toward making more sustainable choices, especially when it comes to optimizing performance, shouldn't be a constant uphill battle. There is compelling evidence that can help them see the light, as we will explore in this chapter.

---

2 Bill Scott, "Improving Netflix Performance", O'Reilly Velocity Conference, June 23, 2008. (http://cdn.oreillystatic.com/en/assets/1/event/7/Improving%20Netflix%20 Performance%20Presentation.pdf)

FIGURE 6-1.
Digital Humvees: image carousels, sharing widgets, video backgrounds; all these components lead to bloated pages and frustrated users. There are better ways.

A Balancing Act

Whereas the website optimization practices discussed in Chapter 5 focus on design tactics, experimentation, and A/B testing to help digital products and services yield better business results, website *performance* optimization (WPO) focuses solely on getting data to users more quickly. In reality, it's a combination of the two that yields the most sustainable results: useless data is still useless no matter how fast it loads and renders on your iPhone. Plus, in the service of sustainability—that is, not compromising the needs of the future—web teams must consider the maintainability of their solutions, as well, as Eric Mikkelsen noted earlier. If it takes an army to make updates—as can be the case with outdated or custom content management systems or those few sites still without a CMS—you're wasting time, money, and frustrating users who expect real-time information in the applications they use. But maintainability has performance overhead ramifications. So let's discuss in more detail how we balance speed, reliability, and sustainability.

WPO DEFINED

At the risk of being too obvious but for the sake of clarity, WPO is defined as a "field of knowledge about increasing the speed in which web pages are downloaded and displayed on the user's browser."[3]

Long considered the realm solely of developers, early WPO efforts focused mainly on optimizing site code and pushing hardware limitations. The best optimization occurs, however, when designers and developers collaborate to bring speedy, reliable *and useful* solutions to users. Leading tech VC Fred Wilson noted in his "10 Golden Principles of Successful Web Apps" presentation that, "There is real empirical evidence that substantiates the fact that speed is more than a feature. It's a requirement."[4] (He put speed at number one on his list.)

---

3 Wikipedia, "Web Performance Optimization". (*https://en.wikipedia.org/wiki/Web_performance_optimization*)

4 Fred Wilson, "The 10 Golden Principles of Successful Web Apps", The Future of Web Apps, Miami 2010. (*https://vimeo.com/10510576*)

So by this logic, everyone involved with bringing a digital product or service to life—designers, developers, product owners, content strategists, project managers, *everyone*—should make optimization their number one priority.

Similarly, according to performance pundit Steve Souders, who coined the term *web performance optimization* in 2004, "80 to 90% of the end-user response time is spent on the frontend. Start there."[5]

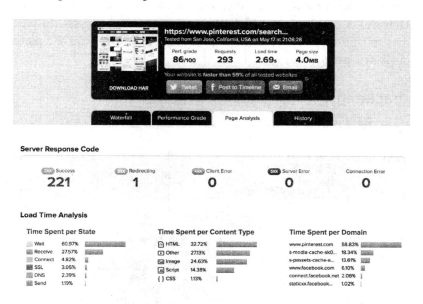

FIGURE 6-2.
The majority of performance hits exist on the frontend

In 2012, Mr. Souders ran performance tests on a range of popular websites and discovered that, across the board, the resources needed to load and execute frontend components—JavaScript, images, CSS/HTML, and other assets, plus page rendering tasks a browser must perform—comprised between 76 to 92% of total page load time. Because design decisions drive so much of what happens on the frontend, designers clearly have a critical role to play in creating optimized solutions. Yet developers are often the ones tasked with optimizing performance.

---

5 Steve Souders, "The Performance Golden Rule", SteveSouders.com, February 10, 2012. (*http://www.stevesouders.com/blog/2012/02/10/the-performance-golden-rule*)

OPTIMIZATION AND COLLABORATION

In a workflow scenario in which designers and developers work in isolation from one another, it's easy for optimization tasks to become lost in the project shuffle. "If a designer creates a page with huge images, no amount of optimization by the developer is going to make that page efficient," says author and professor Pete Markiewicz. "Web designers often don't see their job beyond 'drawing a picture' of a website, which often results in the developer creating bloatware to support their ideas. However, a designer can think of performance at the onset. In fact, they must if they are actually designing for the Web instead of a piece of electric paper."

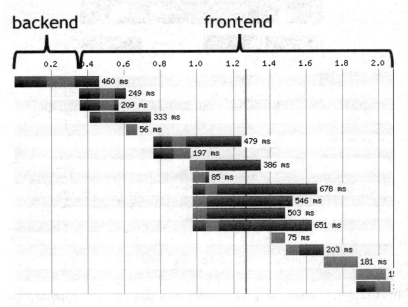

FIGURE 6-3.
The Web isn't just a piece of electric paper—designers must think of performance at the outset

Thankfully, with the rise of Lean/Agile teams and UX as a discipline we have options for maintaining both performance and user happiness across a project life cycle, but it still remains a legitimate concern as we become caught up in client requests, project scope, and deadline-driven decision making. Collaboration and constant communication as well as establishing clear performance goals up front and sticking to those goals throughout a project's life cycle, are key to launching and maintaining more sustainable optimized solutions.

But you still need to answer the question of whether or not you are building a lightweight site that works fast or a maintainable site that can work for years to come. The most sustainable answer to that question will have inevitable trade-offs.

PERFORMANCE VERSUS MAINTENANCE

The fastest product is the one with the leanest code and the fewest assets to load. It puts the right information before users at the time they need it, even if, for instance, those users are on slower, older devices or in an area with limited bandwidth. It knows when to serve barebones content and when the user agent supports a more robust user experience.

But creating the fastest product possible might also mean not using certain libraries or frameworks, ditching a content management system (CMS), or forgoing customer relationship management (CRM) integration. Many things that clients or administrators might want—those familiar UI patterns that shield them from the codey bits—can undermine optimized performance. Conversely, developers might focus only on optimizing code yet leave UI/UX issues alone. Finally, a web app that can't be updated without enlisting the help of a development team (even though it might be crazy fast) is not very future-friendly, which makes it less sustainable.

Developer Eric Mikkelsen on this topic:

> Libraries and frameworks are like web developer pickup trucks. They are helpful tools that make it super easy for me to do my job, but they take up a ton of gas. No one actually wants to get rid of something that helpful when they're told their job is to get things done in a timely manner. So we have to reset what people's jobs are a little. I think this is where a performance budget comes in. Setting a team performance budget tells everyone that it's okay to take extra time and build something that's faster and lighter because that's where our priorities are.

FIGURE 6-4.
Libraries and frameworks are like web developer pickup trucks

These trade-offs are decisions that must be made between project teams and product owners in collaborative settings designed to help achieve consensus. The answers will be different for every project. Consider focusing these discussions on a goal of creating the most sustainable solutions that achieve balance between performance and maintainability while still staying within project specs.

LIBRARIES VERSUS FRAMEWORKS

Who controls the system? This question often drives choices between libraries and frameworks. Or, as blogger and developer Tomas Petricek notes, "A framework defines a structure that you have to fill, while [a] library has some structure that you have to build around."[6] It's up to you to figure out whether the empty house or the collection of individual rooms is more useful. Also, the nature of their structures makes it easy to use multiple libraries in your digital product or service but difficult to use multiple frameworks.

This is relevant to performance optimization and sustainability because both approaches have overhead and can potentially slow down your application. But both speed up the development process, as well. Let's take a look at how they do this.

---

6 Tomas Petricek, "Library patterns: Why Frameworks Are Evil", Tomas Petricek's Blog, March 3, 2015. (*http://tomasp.net/blog/2015/library-frameworks*)

Frameworks

Using a framework as a starting point can save time when developing an application. Web frameworks often alleviate many common programming tasks by providing quick access to commonly used scripts and assets, so developers aren't burdened with programming common features from the ground up each time they start a project.

FIGURE 6-5.
Some common frameworks

When using a framework, your own code sits *within* it. You insert your code into various places inside the framework so that the framework can call upon your code when needed as the application runs. This is useful and saves time, but what happens if you only need a portion of what the framework has to offer? Too bad.

Though using frameworks can save significant development time up front, your application can take a performance hit due to the overhead of the framework itself.

Libraries

When employing libraries, you make calls to external scripts from within your application. These libraries might sit on your own server or, more likely, on a server meant to house shared resources. Google, for example, houses more than a dozen different commonly used JavaScript libraries. Development teams can call upon these libraries to execute common application functions.

FIGURE 6-6.
Some common libraries

Like frameworks, libraries are helpful and time-saving tools, but they come with the added overhead of increased HTTP requests each time a different library is called. However, when using a common resource like Google's Hosted Libraries, if your own site has caching-enabled, it is likely that the end user might already have your script loaded into the browser, thus reducing the overhead of said HTTP requests.

Speed Is Just One Metric

Your application's speed alone can be measured by using tools like Google's PageSpeed Insights, Pingdom Tools, and others. But those tools don't show the complete picture. Developer Dave Rupert addresses this in a blog post titled "The Cost of Frameworks":[7]

> By measuring what can only be measured in terms of page speed means we have no insight to the reasons a framework was employed or how much money it saved the organization. How big is [the] team that built the site? One? Ten? What were past organizational failures that led to adopting a large framework? Did using a framework win out in a lengthy internal Cost-Benefit analysis? Is CSAT up due to swanky animations and autocomplete features that took minutes to build? Was code shipped faster? Was it built by one person over a weekend? Did the abstraction allow the developers to actually have fun building,

---

7 Dave Rupert, "The Cost of Frameworks", DaveRupert.com, November 17, 2015. (*http://daverupert.com/2015/11/framework-cost*)

therefore increasing job satisfaction, therefore reducing organizational churn, therefore reducing cost of the end product for the user? We don't know.

With this in mind, a smart approach might be to start your projects without libraries or frameworks, given their potential overhead, but consider them carefully as you plan and build out your application. A page weight budget, discussed elsewhere in this book, might help, as could a deeper analysis of the tools you're using. As noted previously in this chapter (and as is evidenced by the preceding quote), collaboration and communication are key to weighing which approach seems best for your project.

CMS OPTIMIZATION

Similarly, CMS's bring incredible convenience to the often complicated tasks associated with creating and managing web content. But, if not optimized properly, they can also come with significant overhead. Their extensible architecture brings flexibility to users that frequently comes at the price of performance and reliability. For our purposes here, we will cover WordPress and Drupal, which comprise 52% and 3% of CMS usage, respectively, for the top one million websites.[8]

These systems offer tens of thousands of available plug-ins (WordPress) or modules (Drupal) that handle everything from security to search engine optimization, from commerce to contact forms. Most any web feature you can imagine has likely been turned into a module or plug-in that you can install with just a few clicks. Their simplicity is part of the allure these tools bring to the table. Using them can be very tempting.

But inside these plug-ins and modules lurk some potential pitfalls:

- Each plug-in typically comes with multiple CSS and JavaScript files that might be redundant and can add overhead, affecting performance.

- Updating the CMS or other modules and plug-ins can potentially break your entire website, or take it offline.

---

8   BuiltWith, "CMS Usage Statistics". (*https://trends.builtwith.com/cms*)

FIGURE 6-7.
Get the most out of your CMS by optimizing it for performance

Although the average frontend user might not notice a few extra scripts loading here and there from a single module, every little bit adds up. Install a handful of plug-ins and your site can exhibit performance impediments that *will* be noticed.

The upgrade path mentioned earlier is significantly more detrimental if you're not careful. It is important to keep your software updated because of security concerns. Because these systems are designed for ease-of-use, updating them is understandably meant to be a simple process. In the case of WordPress, an alert shows up in your dashboard whenever an update is available. WordPress is known for simple updates that require just a couple clicks. Because the alert is at the upper-left of your dashboard when you log in, it is the first thing you see, and the temptation to click that update button is palpable. What unwitting content creators don't often realize is that these upgrades can quickly take your site offline. Module and plug-in updates typically lag behind CMS updates, so it is important to ensure that the tools you already use will work together when updated. Otherwise, what seems like a simple process can put you on the fast track to a downed website.

Here are some methods used by developers to streamline these two popular content management systems. It should be noted that the standard web performance optimization practices covered in the following sections can and should also be applied to WordPress or Drupal sites. These recommendations are specific to each CMS.

WordPress

For the purpose of this section, we will focus specifically on two key areas where WordPress can be optimized: plug-ins and themes. We'll also talk a bit about how comments can drag down site performance.

Also, this is by no means a comprehensive list; rather, it represents some of the more common WordPress optimization techniques. There are literally hundreds of tasks you can execute to better optimize your website. Designers and developers worth their salt will also follow guidelines for optimizing frontend, user-facing components, such as those outlined in Chapter 5 as well as CMS-agnostic, general web performance optimization techniques covered later in this chapter.

WordPress Plug-ins. The first thing to do with plug-ins is to deactivate any that aren't being used. Active plug-ins load resources (and make HTTP requests), adding overhead to every page that loads. If a plug-in is not being used, deactivate it.

Keeping in mind that each instance of a plug-in adds its own overhead, here are some specific plug-ins that can help you improve WordPress performance:

Profile performance
If your site is slow, the *P3 Plugin Performance Profiler* will tell you which plug-ins are taking longest to load, how much data they use, and other helpful metrics. Begin there.

Security

As just noted, it's important to keep your site secure and free from malicious hacks. Plug-ins such as *Wordfence, Sucuri Security,* and *All in One WordPress Security and Firewall* can help with this.

Combine scripts

MinQueue combines all your CSS and JavaScript into a single file, which can speed up load times.

Caching

A good caching plug-in like *W3 Total Cache* can speed up page load times by bypassing elements that are already cached in a user's browser.

Optimize the database

A plug-in like *WP-Optimize, Yoast Optimize DB,* or *WP DB Manager,* can reduce the overhead of your site's database by optimizing its elements.

Lazy load images

"Lazy loading" images keeps post images from loading until the user scrolls and they enter a browser's viewport. This speeds up the time it takes for the page to initially load. Davo Hynds, a developer at Mightybytes, created *Lazy Load XT,* a WordPress plug-in that balances configurability with lightweight overhead.[9] Other lazy-load plug-ins include *Lazy Load* and *BJ Lazy Load*.

Compress images

Sure, you compress your photos out of your favorite image editor. A plug-in like *WP Smushit* can add an extra layer of compression by scanning every image you upload to WordPress and stripping hidden, bulky information, reducing file size without losing quality.

Revision control

Left to its own devices, WordPress will save every draft of every post you publish, creating unnecessary information to store in the database. A plug-in like *Revision Control* can help you set a specific

---

[9] http://www.mightybytes.com/blog/lazy-load-xt-wordpress

number of revisions to save. The key is to find a balance between not losing your work (i.e., zero revisions) and cluttering up your database.

Findable content
A plug-in like *Yoast SEO* can help you create pages that meet SEO standards, allowing your content to be found more quickly.

Creating mobile pages
PageFrog's plug-in for Google AMP and Facebook Instant Pages can autogenerate stripped-down versions of your page that load up to 10 times faster than the original.

These are just some of the helpful plug-ins that can help optimize your WordPress site. Just remember, more plug-ins = more potential overhead, so choose wisely.

Your WordPress theme. The concept of progressive enhancement (covered in Chapter 5) applies here. A performance-optimized and more sustainable WordPress theme would embrace this concept—that is, display base content to browsers with few supported features and then layer on additional functionality as a user agent allows. Premium themes such as the *Thesis Theme Framework* or *Lucid* are feature-rich and responsive without slowing your site down. But there are many others, as well.

A good theme also uses valid HTML, which loads quickly across browsers and devices. Unfortunately, some themes can contain spam or malware. You can check the validity of the theme you're using with plug-ins like the *WordPress Authenticity Checker* or *Exploit Scanner*. You can also check whether the base code is standards-compliant by using the World Wide Web Consortium (W3C)'s Markup Validation Service.[10]

Next, check the assets your theme uses:

Number of queries
Can you hardcode static elements into the theme? This will reduce the number of HTTP requests. This could be relevant for static menus, site title, and so on.

---

10 *https://validator.w3.org*

Images
>Are there unnecessary images? Are those that are being used properly compressed? Are they in the right format (JPG, GIF, PNG, etc.)?

Number of files
>Can you reduce the number of files needed to display pages—CSS, JavaScript, and so on—by combining them into a single, optimized file that is minified? (This will likely change with wider adoption of HTTP/2.)

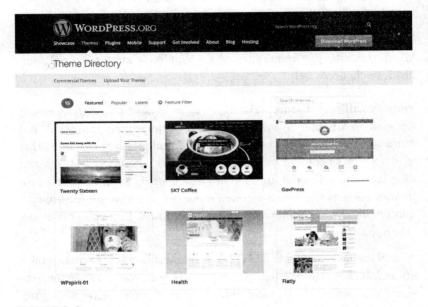

FIGURE 6-8.
There are tens of thousands of WordPress themes out there; if you're not creating a custom theme, be sure to go with one that is feature-rich yet still lightweight

Pete Markiewicz also notes: "I think the WordPress JavaScript API will change the game for theme design, allowing pages to be vastly streamlined. If possible, designers might want to consider using themes with the JavaScript API over older PHP-driven themes."

Of course, every site has different needs and requirements, so what works for one WordPress site might not work for another. Keeping the aforementioned considerations in mind when creating and editing your theme can help reduce overhead and speed up load times.

Comments, pingbacks, and trackbacks. Disabling pingback and trackback functionality in WordPress can streamline the amount of work your site does tracking who's mentioned your posts. People can still link to you. Those links just are not added to your WordPress database. Given how much spam is out there, this could potentially make a significant difference in performance. Automated spam comments can begin to eat up your MySQL database. As the folks at SEOChat note, "If you continually ignore this, you might notice that your site is down, or you won't be able to post because your MySQL disk space has exceeded its quota."[11]

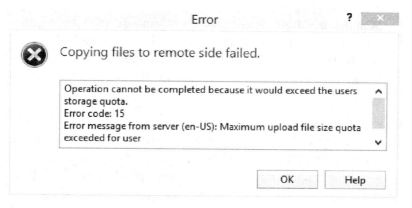

FIGURE 6-9.
Spam eats database for breakfast. Mmmmm.

If your site doesn't rely heavily on user comments, you might consider disabling commenting functionality altogether. Monitoring, moderating, and interacting with comments (and the spammers who create them) can be a resource-intensive process. We will cover comments in more detail later in this chapter.

---

11 Seochat, "Prevent Comment Spam from Damaging Your WordPress Website". (https://www.seochat.com/c/a/search-engine-optimization-help/prevent-comment-spam-from-damaging-your-wordpress-website)

Drupal

With more than one million websites (as of this writing), Drupal runs about 3% of the Internet.[12] As with WordPress, the techniques outlined in the following sections are not meant to serve as a comprehensive resource but rather as a place to get started when thinking about optimizing your Drupal site's performance.

Your Drupal modules. Keep in mind that, like WordPress, unnecessary and unused Drupal modules add overhead to page loads. When considering a module strategy, be sure each one you choose will be used and delete or disable any that won't.

Remove unnecessary HTML
The Fences module creates leaner markup and removes extra classes to keep your code lightweight.

Compress and aggregate files
Modules like Minify, Speedy, and the Advanced CSS/JS Aggregation module will help you compress files and, in the case of the latter, run JavaScript in the footer (to stop page render blocks), and use Google's shared jQuery library, which will reduce HTTP requests if cached in the user's browser.

Serve optimized images
Though you should always compress images before uploading them to the CMS, an added boost from the ImageCache module, which optimizes images *after* they are uploaded, might help, as well.

Load lazily
Loading images as the user needs them can help offset a large download up front. The Lazy Loader module will load images as they come into the browser's viewport.

Findable content
A module like the SEO Checklist can help you create pages that meet SEO standards, making it possible for your content to be found more quickly.

---

12 Mike Gifford, "Tips for a Sustainable Drupal 7 & 8 Website", OpenConcept Consulting, October 23, 2015. (*http://openconcept.ca/blog/mike/tips-sustainable-drupal-7-8-website*)

Speedier databases
: Your site's database can become bloated and slow down performance. The DB Maintenance module can help optimize database files for speedier delivery of content.

Enable caching
: Drupal can also cache your files using modules like Memcached or Varnish, saving your browser from making repeated server trips for files that are already cached.

Share resources
: By using shared libraries or placing your static content on a content delivery network (CDN), you reduce the amount of time it takes users to get content either through proximity to the server or via user-cached files. The CDN and/or Boost modules can help with this.

Search on site
: Adding search functionality to your own site via Drupal's core search functionality or the Apache Solr module can help users find what they need more quickly, saving time and energy.

Your Drupal theme. Similarly, here are two things you can do to keep your Drupal theme optimized:

Start smart
: Begin with solid base theme like Zen or Adaptive Theme, which are built with modern web standards in mind: they are responsive, accessibility-compliant, and use semantic HTML5 and CSS 3.

Remove that registry
: Drupal's theme registry holds cached theming data which is helpful when adding theme hooks or new modules, but unnecessary when a site is live in production. Disabling the theme registry on a production site can help increase performance.

Performance Rules

Google, Yahoo, and many others have published extensively on guidelines for optimizing website performance. Google's PageSpeed Insights tool, which grades your website based on speed, publishes a set of rules it has devised for grading criteria. Yahoo's Exceptional Performance team has created similar recommendations. The following lists outline some common tactics.

Yahoo includes the following rules as part of its "Best Practices for Speeding Up Your Web Site":[13]

- Minimize HTTP requests by reducing the number of objects—scripts, images, stylesheets, and so on—required for a browser to render a page. You can do this by using CSS sprites, image maps, inline images, combining scripts, and so on.

- Use a CDN to deploy your content across multiple servers so that proximity to request yields faster page loads. This can potentially be cost-prohibitive for organizations with constrained resources like startups or some nonprofits, but these services can yield dramatic results. Yahoo reported improved end-user response times by up to 20% with this approach.

- Adding an Expires or Cache-Control header could speed up pages for users by caching commonly used scripts, images, or other components. As users navigate from screen to screen, common page elements are already preloaded.

- Compress site components with Gzip to reduce their file size by about 70%. HTML, stylesheets, and scripts can all benefit from this server-side compression technology that can reduce page weight and speed up a user's experience.

- Put stylesheets in the document HEAD to render them progressively, which gives the appearance of faster loading.

- Put scripts at the bottom of your document to accommodate parallel downloads and load your page faster. Scripts block parallel downloads, so putting them at the bottom allows other page components to start loading first.

---

13 *https://developer.yahoo.com/performance/rules.html*

- Avoid CSS Expressions, which are evaluated by the browser when the page is rendered, resized, scrolled, or when a user moves her mouse. This can significantly slow down page performance.

- Make JavaScript and CSS external so that they can be cached by the user's browser. Even though this initially increases HTTP requests upon first load, after it's cached those external files don't need to be downloaded again, which in turn decreases HTTP requests on subsequent pages. Conversely, inline JavaScript and CSS are downloaded every time a page loads, which increases the size of your HTML documents.

- Reduce DNS lookups by reducing unique hostnames included in your pages. This includes those used in URLs, images, script files, stylesheets, embedded objects, and so on. Each DNS lookup can add anywhere from 20 to 120 ms in load time to your page.

- Minifying JavaScript and CSS will remove excess unnecessary characters, comments, and white spaces from your code to improve load times. Tools such as JSMin and YUI Compressor can take care of this for you.

- Avoid redirects when possible because they can slow down the user's experience. If you must redirect from a page that no longer exists to a new one, use a 301 permanent redirect to avoid any potential SEO penalties. Not including trailing slashes (/) at the end of your URLs forces autoredirects by the server unless you set it up otherwise.

- Remove duplicate scripts in any page. Why load the same script twice or more while also increasing HTTP requests from your pages?

- Configure Entity Tags (ETags) to validate and match cached scripts to those being requested by the server. ETags work well for sites on the same server, but don't typically translate from one server to another due to server-specific embedded data, which reduces their usefulness.

- Make Ajax cacheable by using the Expires or Cache Control methods noted earlier to speed up applications that might require lengthy asynchronous JavaScript and XML responses.

- Flush the buffer early using PHP's flush() function so that partially ready HTML documents can begin fetching data without waiting for the backend server to stitch together your pages.

- Use GET for Ajax Requests rather than POST when using XMLHttpRequest, as the latter implements a two-step browser process in contrast to GET's one (unless you have a lot of cookies).

- Post-load components that aren't absolutely required to render the page initially. This could include content that requires scrolling to view, JavaScript animations or drag-and-drop behaviors, hidden content, and so on.

- Preload components to take advantage of when the browser is idle to request components (images, styles, scripts, etc.) that will be needed in the future. Subsequent pages will in turn load much faster.

- Reduce the number of Document Object Model (DOM) elements in your pages to speed up access by scripting languages like JavaScript. More complex pages can slow down DOM access.

- Split components across domains to maximize parallel download opportunities. Just don't use more than two to four domains, for what you gain in parallel downloads you could lose again in DNS lookup time.

- Minimize the number of iframes in your documents to reduce data download costs that occur by using them. iframes can be helpful with slow third-party content like badges and ads, but even a blank iframe can add to your page weight.

- Avoid 404 errors, which use HTTP requests while returning no valuable information to users.

- Reduce cookie size and eliminate unnecessary cookies to minimize their impact on user response time.

- Use cookie-free domains for static components, such as images, that don't require cookie requests. Static components can be served from a cookie-free subdomain, for example, to minimize unnecessary network traffic.

- Minimize DOM access by JavaScript, which can make pages less responsive. Cache references to accessed elements, update nodes "offline" and then add them to the tree, and don't fix layout issues with JavaScript.

- Develop smart event handlers that can delegate events in order to manage how often and when handlers are executed. Handlers that are executed too often can make pages less responsive.

- Optimize images to ensure that they preserve best possible image quality while maintaining smallest possible file size. (Chapter 5 covers this in detail.)

- Optimize CSS sprites to reduce file size while maintaining image quality. (Also covered in detail in Chapter 5.)

- Don't scale images in HTML; instead, create them at the size they're needed.

- Make favicon.ico small and cacheable because a browser will always request it (and send a cookie in return) whether you want one or not. If possible, make the image 1 KB or smaller and set a lengthy Expires header so it is cached.

- Keep components under 25 KB because iPhones don't cache anything larger. Note this is *uncompressed* size.

- Pack components into a multipart document to reduce HTTP requests.

- Avoid empty `<image src>` tags in HTML and JavaScript because they still make server requests even though they are empty.

In addition to these recommendations, Google's PageSpeed Insights rules include the following:[14]

Speed:

- Improve server response time to less than 200 ms to avoid performance bottlenecks. Slow server times could have numerous potential causes, including slow application logic, slow database

---

14 *https://developers.google.com/speed/docs/insights/rules?hl=en*

queries, CPU resource or memory starvation, and so on. Many automated web application monitoring solutions are available to track performance.

- Inline CSS, if the scripts are small. This lets the browser render the styles individually rather than waiting for an entire stylesheet to load. Just make sure you don't duplicate scripts.

- Prioritize visible content to load above-the-fold content first by structuring your HTML and CSS accordingly, and also reducing its size.

Usability:

- Avoid plug-ins that can cause crashes, hangs, or security issues. Even though these plug-ins help a browser process special kinds of content, most are not supported on mobile devices.

- Configure a viewport to control how a page is rendered on various devices to provide optimal viewing for users.

- Size content to fit the viewport, once configured.

- Size tap targets appropriately for the devices your content will be viewed on. If buttons or form fields are too close together, they can be difficult for users to accurately use on touchscreen devices.

- Use legible font sizes that are appropriate for the devices defined in your viewport.

The preceding list is by no means comprehensive, but reflects common recommendations for speeding up your site. As we will see in the sections that follow, web teams devise all sorts of methodologies for hitting performance goals on a web project. And speed isn't the only metric.

Speed, Reliability, and Version Control

Where a common performance quandary arises is in providing the best experience for users while also supporting as many devices and platforms as possible. In sustainability parlance, this goes back again to the concept of serving the needs of the present without compromising the needs of the future—but in reverse. Here we're trying to support the devices of the past while providing the best possible experience for users with the latest browsers, quickly and efficiently, of course. Progressive enhancement and all that.

This approach will provide a more reliable experience that works across a wider array of devices and platforms, but it could come at the cost of speed. To address this conundrum, following are several approaches that Eric Mikkelsen and the rest of the Mightybytes development team have come to rely on.

AUTOPREFIXER

With Autoprefixer, you can write CSS for today while also supporting older browsers. Vendor prefixes allow browsers to support more experimental CSS declarations. You can use some of these declarations to create more efficient and user-friendly CSS layouts. This is great news for users of the latest and greatest browsers, but not so great for those who use older browsers or browsers with a different interpretation of CSS declarations than what you wrote.

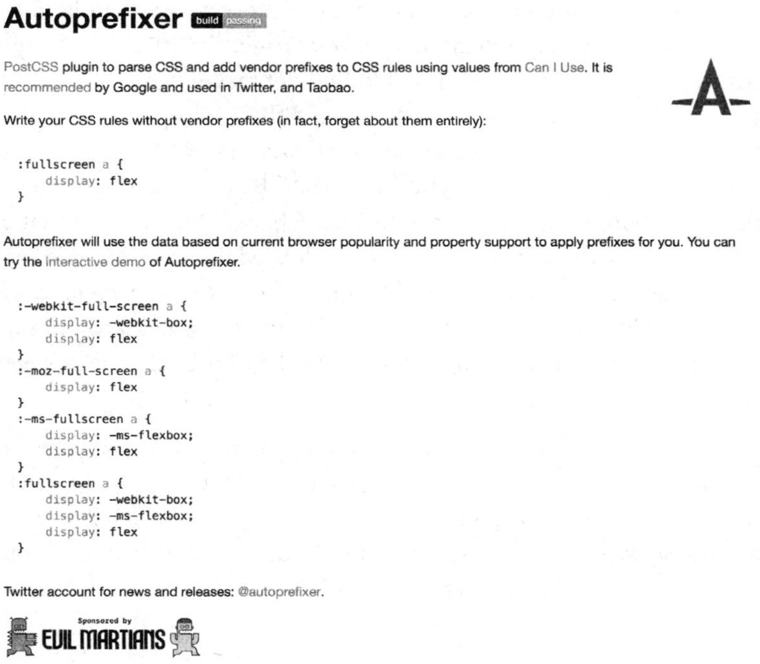

Autoprefixer build passing

PostCSS plugin to parse CSS and add vendor prefixes to CSS rules using values from Can I Use. It is recommended by Google and used in Twitter, and Taobao.

Write your CSS rules without vendor prefixes (in fact, forget about them entirely):

```
:fullscreen a {
    display: flex
}
```

Autoprefixer will use the data based on current browser popularity and property support to apply prefixes for you. You can try the interactive demo of Autoprefixer.

```
:-webkit-full-screen a {
    display: -webkit-box;
    display: flex
}
:-moz-full-screen a {
    display: flex
}
:-ms-fullscreen a {
    display: -ms-flexbox;
    display: flex
}
:fullscreen a {
    display: -webkit-box;
    display: -ms-flexbox;
    display: flex
}
```

Twitter account for news and releases: @autoprefixer.

Sponsored by EVIL MARTIANS

FIGURE 6-10.
Autoprefixer supports newer CSS styles while also supporting older browsers

With Autoprefixer, developers have the ability to set how many versions back they want to support. It uses information housed in the "Can I use" database to control which declarations are executed and which aren't based on the user's browser profile. It eases the process of phasing out code to support older browsers, providing a useful experience across browsers and devices.

Similarly, when your page boasts a feature like background gradients or rounded corners, you can rely on Autoprefixer to support those in current browsers while not serving the fancier styles to older browsers. A box with text in it, for example, still reads just fine without rounded corners. But when you have more complex features like Flexbox or Grid Layouts, you need more complex fallbacks than Autoprefixer can offer. By allowing frontend developers to use variables and *mixins* in their CSS, they can easily remove these fallbacks as browser support increases with time, not requiring full overhauls for speed increases.

SHARETHIS, ADDTHIS, DUMPTHIS

Social media sharing plug-ins, those ubiquitous buttons that beg you to share content across your social channels every time you land on a blog post, can add significant heft to your pages. In 2013, UX designer James Christie ran an experiment to find out just how much these widgets add to a typical page.[15] He discovered that adding four social widgets to a web page also added an additional 64 HTTP requests and caused his page to bloat from 80 KB to 480 KB. The widget-addled page also took more than six seconds to load. Christie took his calculations further to note that if one million people loaded the page it would result in 1,727 hours, or 71 days, of collectively wasted time waiting for data to load. In addition, those 71 days also added up to 379.8 GB of extra data sent and, by his calculations, about 7.41 tons of greenhouse gas emissions, about the same amount as four transatlantic flights.

Still, your marketing department wants people to share your content and they don't trust people's abilities to copy and paste a URL to Facebook or Twitter. What are you to do? Some sites have taken to simply adding hypertext links to their preferred social networks on content pages. Rather than making unnecessary additional HTTP requests

---

15 JC UX, "Social Sharing Buttons: Page Weight Experiment". (*http://jcux.co.uk/oldsite/posts/buttons.html*)

and automatically loading the widget code and images for each individual network upon page load, this approach puts the decision in a user's hands. The result is a much lighter page weight that only loads sharing data when a user selects the network with which they want to share content.

Or, you can trust that users know how to copy and paste a URL.

FIGURE 6-11.
Social media sharing plug-ins can add significant heft to your pages (not to mention a lot of extra HTTP requests)

COMMENTS AND PAGE BLOAT

A similar situation can arise with blog comments. This can be especially problematic if you use a third-party comments system like Disqus, Livefyre, and so on. In tests we ran on the Mightybytes blog, our comments system added an additional 14 HTTP requests, 34 MySQL queries, and .4 MB to each page that used it. This also translated into about a half-second of additional load time, or nearly 69.5 collective hours and 400 additional unnecessary gigabytes were one million people to download a page from our site that included the comments system. This would be slightly more greenhouse gas emissions than the four transatlantic flights noted previously in James Christie's calculations.

All this said, comments certainly have their place. I'm not advocating that every blog remove comments to save the planet. Comments give users a way to carry on content-centered conversations, share ideas (and links), and create connections. For some sites or products, this is absolutely necessary. For other blogs, the comments section unwittingly exists as a repository for spam and little else. You have to weigh the choices between the value of comments versus the overhead they add to your site.

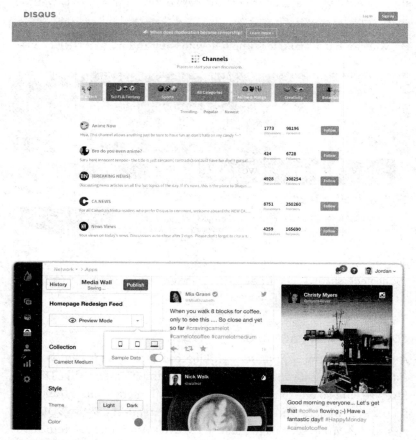

FIGURE 6-12.
Third-party comments systems can really slow things down

HTTP/2 AND PERFORMANCE

Finally, another area where performance could improve significantly is with HTTP/2, which promises to speed up access to content by allowing simultaneous downloads. In early Internet days, when web pages were much simpler, loading them didn't require as many data requests. Today's web pages are more resource-intensive with shared libraries, more images, embedded videos, JavaScript, CSS, and so on.

Web teams have used techniques like CSS sprites, shared libraries, and caching to reduce the number of server requests on a page for many years, but these are just workarounds for what is an inherent problem with HTTP/1: it allows only one outstanding request per TCP connection. HTTP/2 is an attempt to modify this rule, and for performance-focused developers, this means a lot of standard practices are about to get thrown out the window.

HTTP/2 supports compressing headers and cookies before sending them to the user. Other improvements, like multiplexing using a single TCP connection and server push, could affect the process by which developers optimize their code. With wider adoption, web teams will no doubt discover new ways to optimize performance, many of which will likely affect developer workflows.

HTTP/2 requires support on both the server and browser sides to function properly. As of this writing, caniuse.com reports that 70.8% of global browsers fully support (63%) or at least partially support (7.18%) HTTP/2.[16] That includes Chrome, Edge, Firefox, and Opera, with Safari following in 2016 as soon as Safari 9 is released. Adoption is happening quickly on the server side, as well. The HTTP/2 Wiki on GitHub has an up-to-date list of server-side adoptions.[17]

Because HTTP/2 is backward-compatible with HTTP/1.1, you could ostensibly ignore it and everything will continue to work. From a performance perspective, there are, however, a few things you should know.

To update your own website, two important things need to happen first:

- Server software will need to be updated to versions that support the protocol.
- You will also need a Secure Sockets Layer (SSL) certificate for your site to ensure secure connections. You should do this anyway, as Google identified HTTPS as a ranking signal in August 2014. Secure sites rank higher.

---

16 Can I use, "Can I use... Support tables for HTML5, CSS3, etc". (*http://caniuse.com/#search=http2*)

17 *https://github.com/http2/http2-spec/wiki/Implementations*

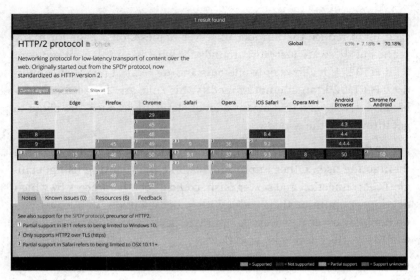

FIGURE 6-13.
HTTP/2 will make some more sustainable workarounds—like CSS sprites and aggregating scripts—unnecessary

In some instances, workarounds discussed elsewhere in this book for HTTP/1.1 shortcomings will actually become less efficient:

Concatenating CSS and JavaScript files
   HTTP/2's server push feature will deliver all of your separate CSS files directly to the user without them having to make individual requests for each one. Not only will it figure out which files the user needs, but it can deliver those files faster than a single concatenated script file could compile and deliver them via HTTP/1.1.

Embedded image data
   Base64-encoded images embedded with CSS will reduce HTTP requests, as is discussed in Chapter 5. This workaround will increase the size of your CSS files. Because multiple HTTP requests aren't a concern with HTTP/2, visitors will need to download all of this data regardless of whether they visit the pages that use that data. So, although this workaround can improve performance for sites that use HTTP/1.1, it could slow performance for others.

Sharding

> HTTP/1.1 restricts the number of open connections between your website and server. Historically, a workaround was to retrieve assets from multiple domains by using a technique called domain sharding. You can request as many resources as you need with HTTP/2, eliminating the need for sharding.

Sprites

> HTTP/2 supports multiplexing, so resources are no longer queued, which can result in loading lag time. Chapter 5 discusses using CSS sprites as a workaround for this, which is smart for HTTP/1.1 but unnecessary for HTTP/2.

If you don't have control over your hosting environment, you will need to wait until servers are updated to support serving pages via HTTP/2. Similarly, when the majority of your users run browsers that support the protocol, you should consider switching, but maybe not before. Analytics data can tell you this. That said, SSL certificates are readily available from many sources. That is something you can, and should, do today.

Finally, for the aforementioned reasons, HTTP/2 will change your regular design and development workflow, so you should assess your own needs as well as those of your team and end users to ascertain a timeline that works for all involved to make the change.

Workflow Tips

We have talked a lot about techniques in this chapter so far, but how are developer workflows affected by a web team that is driven by performance?

LEAN/AGILE WORKFLOWS

We covered Lean/Agile workflows earlier in the book, but it is worth mentioning again in terms of how, specifically, these workflows can address performance issues:

- Iterative testing in every sprint will often reveal performance hiccups that can be caught early on rather than in beta, or worse, post-launch.
- Agile teams can build performance-specific sprints into their process.

- User testing is typically built into Agile workflows, which lets teams identify performance impediments early and fix them within the course of a sprint.
- You can set page "budgets" in a sprint planning meeting to set ongoing guidelines for the project that all teams must adhere to.

STANDARDS-BASED DEVELOPMENT

Using web development techniques based on accepted standards put forth by the W3C will always yield more sustainable results. By the time standards are approved, functionality associated with those standards is typically well entrenched within current browser capabilities. Creating applications based on standards will ensure a wider audience. Testing against those standards will ensure your information can be accessed across the widest array of devices and platforms.

The adoption of standards can be a slow process, however. Standards move at a much slower pace than the release of new devices, which can make for many design and development challenges.

By the time HTML5 and CSS 3 were officially approved as web standards—a process that took several years—nearly all modern browsers had already long supported them.

FIGURE 6-14.
Browsers supported HTML5 and CSS 3 long before they officially became standards

That said, keep in mind that these efforts are for naught if you don't provide solid value for both businesses and users, or if your content sucks:

> To manage in a world of ever-increasing device complexity, we need to focus on what matters most to our customers and businesses. Not by building lowest common-denominator solutions but by creating meaningful content and services. People are also increasingly tired of excessive noise and finding ways to simplify things for themselves. Focus your service before your customers and increasing diversity do it for you.[18]

So ground your work in performance-driven standards but ensure that there is a solid business foundation that includes real value propositions for users, as well.

VALIDATING YOUR WORK

Not sure if your digital product or service is as optimized for performance as it could be? Here are some tools that can help.

Use validators like Google's PageSpeed Insights or Pingdom Tools. These services will guide you through the optimization process, offering helpful tips and suggestions for maximizing performance.

Similarly, the W3C's Validation Markup Service can help you create digital products and services that meet the rigorous standards set forth by the W3C.

508 Checker, by the folks at FormStack, can help you get a better handle on how your site meets accessibility standards for people with disabilities.

Finally, a tool like Ecograder or CO_2 Stats (covered in detail in Chapter 7) can help you to better understand how poor performance can lead to wasted energy and a larger environmental footprint for your application.

---

18 *http://futurefriendlyweb.com/thinking.html*

FIGURE 6-15.
There are easy ways to validate your work with web-based tools

Accessibility and Sustainability

Whether it's energy, water, sanitation, or something else, universal access is a key part of all sustainability frameworks. An accessible website gives people with disabilities, who might use enabling devices such as screen readers, the ability to experience digital products or services in meaningful ways. If we look at accessibility through the lens of sustainability we must take a broader approach. The French Open Web Group states that accessibility is good for the planet:[19]

> When working on Web accessibility, one finds frequently that some practices result in reducing the size of pages, or the amount of data transferred to the user. Furthermore, an accessible Web site is generally simpler and therefore faster to view, not only for users with disabilities.
>
> Every small earnings, effortlessly obtained, add to each other, and humbly contribute to the global effort. Much rather like those soda cans you throw in the right recycling bin: this is not what will fundamentally change things; but things will not change unless we do at least our share.

Open Concept Consulting's Mike Gifford agrees:

> In a fast-paced world, having well-structured content that is written in plain language helps people get what they need and act on it. Numbers vary, but a sizable portion of the population has a disability that can affect how they use the Web. The Web Content Accessibility Guidelines (WCAG) push for sites to be perceivable, operable, understandable and robust.

If you look at the entire range of potential user disabilities—low vision, color-blindness, mobility challenges, debilitating injuries, and so on—the numbers could perhaps reach up to 20% of any given population. What is the impact of either excluding these people or simply having to service them in a different way? Thus, following accessibility guidelines just makes sense.

---

19 OpenWeb, "Accessibility Is Good for the Planet". (*http://openweb.eu.org/articles/accessibility-is-good-for-the-planet*)

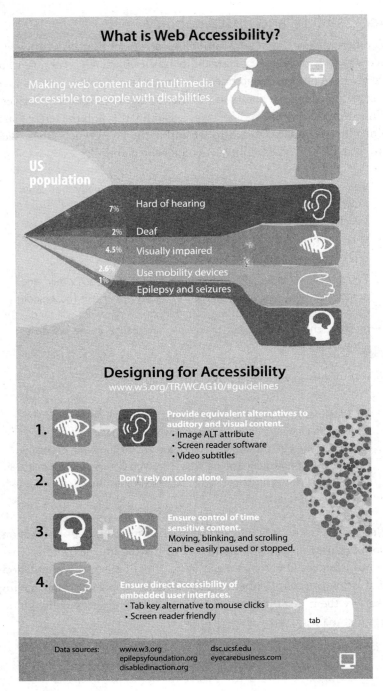

FIGURE 6-16.
Accessibility is a key component to all sustainability frameworks

Complicating this is the need to optimize digital products and services for new browsers and devices that appear around every corner while also supporting older or less popular models with a more basic experience. Though it's not always possible to create the best experience for all users on all devices, we do have a responsibility to build something that's useful for everyone who interacts with our product or service. This can cause an ever-dizzying array of contingencies that must be planned for, adding overhead not just to the product or service itself but also to the process of its creation.

Brad Frost, in his post "Support vs. Optimization," states:[20]

> The minute we start giving the middle finger to these other platforms, devices, and browsers is the minute where the concept of the Web starts to erode. Because now it's not about universal access to information, knowledge and interactivity. It's about catering to the best of breed and leaving everyone else in the cold. All of a sudden, the "native vs web" argument is no longer bullshit. It becomes more an apples-to-apples comparison, where web experiences only work on the platforms that happen to provide thousands of gorgeous native apps.

Potential Barriers and Workarounds

There are some hurdles to reaching a lean, mean Internet. Focus on performance is increasing but still not commonplace. Few web teams, for instance, allot a performance budget. Defining and implementing good performance practices so that they become a rote part of any process will ultimately lead to more accessible and sustainable digital products and services. But this means getting the entire Internet supply chain—product managers, clients, designers, developers, data centers, agencies, and so on—to focus on performance and efficiency. There is still a long way to go.

Then there is the curious case of the DIY website. Like them or not, tools such as Wix and Squarespace make it easy for many people without web design experience to create sites. This leads to just as many poorly designed web solutions. Jevons Paradox at work again. Inexperienced

---

20 Brad Frost, "Support vs. Optimization". (*http://bradfrost.com/blog/mobile/support-vs-optimization*)

site creators make bad decisions that undermine user experience and decrease performance. Plus, some of these solutions are closed source, which could mean less accessibility, less efficiency, and more wasted energy.

Standard APIs, hosted libraries, and frameworks offer easier access to more features, which makes more-robust applications possible. They can also significantly increase the amount of data transferred from page to server, and in the case of frameworks can contribute to applications that are larger than they need to be. This is one of those situations where the advantage of shared data might outweigh the barrier of slower applications, but tell that to the guy in rural India who's trying to find directions on your travel app.

And of course, it's the Internet. Standards always shift and that's typically a good thing, but shifting standards means those who don't keep up are potentially creating subpar products and services that could perform poorly.

Conclusion

Here's what we covered in this chapter:

- Why optimizing performance is a critical web sustainability component.
- Techniques for creating better optimized digital products and services.
- Workflow tips for assessing performance.

Hopefully the processes outlined herein will help you create fast, reliable, performance-optimized solutions that are accessible to as many people as possible.

Action Items

Try these three things:

- Run your site, product, or service through the speed, accessibility, and standards validators mentioned in this chapter. Is there room for improvement?
- Create a checklist of tasks you can do to make your website or app more accessible, standards-based, and perform better.
- Start attacking that list!

[7]

Digital Carbon Footprints

What You Will Learn in This Chapter
Building a more sustainable website is great, but can you actually calculate its emissions? This chapter covers why that's quite a challenge, with many moving parts.

Estimating a Carbon Footprint
What is the carbon footprint of a website or mobile app? It's a question I have been trying to find an answer to for several years. My holy grail has always been to find a simple formula that could be applied to a digital product or service to estimate its environmental footprint. Turns out, someone actually owns a patent for that very thing.

More on that in a bit.

FIGURE 7-1.
The total amount of CO_2e produced by your every day activities makes up your carbon footprint

Let's first take a step back. Everything we do produces some measure of waste: eating, driving, working, even breathing. Much of this waste either creates or consists of greenhouse gases (remember CO_2e from Chapter 1?). To decipher the level of impact our activities have on the environment, it becomes important to measure, or at least accurately estimate, the amount of greenhouse gases they produce.

Writing for *Carbon Management*, authors Laurence A. Wright, Simon Kemp, and Ian Williams define a carbon footprint as:[1]

> A measure of the total amount of carbon dioxide (CO_2) and methane (CH4) emissions of a defined population, system or activity, considering all relevant sources, sinks and storage within the spatial and temporal boundary of the population, system or activity of interest. Calculated as carbon dioxide equivalent (CO_2e) using the relevant 100-year global warming potential (GWP100).

According to Nathan Shedroff, author of *Design Is the Problem*:

> A carbon footprint is a way of estimating the amount of carbon dioxide our activities generate, and by understanding this, we can find ways to lower these emissions.[2] It represents the total amount of carbon dioxide our activities generate—from heating our homes to driving cars to eating and drinking to working and living. Carbon footprints are difficult to calculate exactly because there are so many variables. However, most carbon footprint calculators do a great job of estimating our personal or corporate carbon emissions by using averages.

In their book *Ecological Economics Research Trends*, authors Thomas Wiedmann and Jan Minx offer what is, in my opinion, the most concise definition:[3]

---

1 Laurence A. Wright, Simon Kemp, and Ian Williams, "'Carbon Footprinting': Towards a Universally Accepted Definition", Carbon Management 2:1 (2011): 61–72) (http://www.tandfonline.com/doi/abs/10.4155/cmt.10.39)

2 Nathan Shedroff, *Design Is the Problem* (Brooklyn, NY: Rosenfeld Media, 2011). (http://rosenfeldmedia.com/books/design-is-the-problem)

3 Thomas Wiedmann and Jan Minx, "A Definition of 'Carbon Footprint'" in *Ecological Economics Research Trends* (New York: Nova Science Publishers, 2007), 5.

> The carbon footprint is a measure of the exclusive total amount of carbon dioxide emissions that is directly and indirectly caused by an activity or is accumulated over the life stages of a product.

Their definition can be broken down to include the activities of individual people, populations, governments, companies, organizations, processes, industry sectors, and so on with consideration not only on direct emissions (such as those occurring presently with internal materials or processes), but also indirect emissions (those that are external and upstream or downstream in the life cycle).

CALCULATION CRITERIA

Remember the virtual life cycle proposed by Pete Markiewicz that we discussed in Chapter 1? This could be a jumping-off point for determining where digital products and services use or influence sustainability principles over the course of their lifetimes.

TABLE 7-1. Can life cycle assessment principles be applied to virtual properties?

LIFE CYCLE ASSESSMENT	VIRTUAL LIFE CYCLE ASSESSMENT
Materials	Software and visual assets
Manufacturing	Design and development
Packaging	Uploaded to the Internet
Distribution	Downloaded through the network
Usage	Interaction, user experience, completing tasks
Disposal	Data erased from client

It's important to remember that impacts are harder to measure in retrospect, so keeping the virtual life cycle assessment (VLCA) components in mind as you devise, design, and develop will make estimating impacts easier than trying to decipher them when you're all done. For digital products and services, this would best be taken on by a product manager versus a designer, developer, or project manager.

Here are some questions to ask at the outset:

- How many workstations and devices are you using throughout the process? What are their electricity needs?

- How long are your computers running during design and development? (If you already track your time, this can give you a rough estimate.)

- How big are your source files? Do they live on an internal server or in the cloud?

- Similarly, how big are the files you upload to your server? What's the total amount of data?

- How many users on average do you have per day? How much data do they download? How long do they stay? Where do they come from?

- What devices do they use? What are the power requirements of those devices?

- How many unused files are taking up space on your servers? How much time/effort/electricity does it take to delete them?

FIGURE 7-2.
How do our every day digital activities fit into a carbon footprint?

Of course, it's tough to estimate some of the aforementioned components. Adobe Photoshop, for instance, is 26 years old as of this writing. That's a long life cycle. How many hours and resources went into making that thing what it is today? Hard to know.

Pete Markiewicz offers some thoughts on VLCA's usefulness in estimating the footprint of a digital product or service:

> Computing carbon footprints in a hardcore faction requires considering the local site's energy use—along with the energy use by external features needed to keep the site going, or "externalities." As an analogy, take a hydrogen-powered car. The car itself is highly efficient and hardly pollutes. But at present, hydrogen is created via a steam process which releases lots of CO_2. If you also factor in hydrogen loss from the fuel tanks over time and the cost to build a hydrogen infrastructure, the next is more polluting than our current fossil fuel system.
>
> Deep analysis requires including the site, along with the externalities be subjected to an LCA, or Life Cycle Assessment. In other words, you compute the energy footprint of the site, its external resources (e.g., a team of content creators), and sum it over the expected lifetime of the site. Compared to physical projects, this is open-ended but most sites go several years between major design revisions which might provide a stopping point.
>
> In practice, one should do an 80/20 approximation of an LCA, getting a reasonable energy budget for:
>
> - Embodied energy of creating the website (including keeping the lights on in the design shop).
> - Maintenance energy of creating the website (including content maintenance and IT).
> - Site delivery energy (traditional WPO and web footprint).
> - Site inclusiveness (how many people could use the site can actually use it).
> - For international sites, one might use the Web Index (http://webindex.org) to adjust the score, based on the value of the Internet to the target audience by country."

The VLCA seems to be the most comprehensive approach, but the inventory analysis of a VLCA is also potentially the most complicated to accurately estimate, given the conditions just noted.

Proposing a Framework

For this section, I asked interviewees featured in other sections of this book to outline a potential framework or approach they might use to accurately estimate the carbon footprint of a digital product or service. Here are their answers.

In a 2013 article titled "Sustainable Web Design" on A List Apart, James Christie proposed a framework for estimating the carbon footprint of a website:[4]

- A 2008 paper from the Lawrence Berkeley National Laboratory suggests it takes 13 kWh to transmit 1 GB.[5]
- According to EPA figures, the average power plant in the United States emits 1.2 pounds of carbon dioxide equivalent (called CO_2e) per kWh produced (other countries have higher or lower averages depending on their energy policy).[6]
- If we multiply 13 kWh by 1.2 pounds, we get 15.6 pounds of CO_2e—and that's just to transfer 1 GB of data.
- If one million users each download a typical page, which now averages 1.4 MB, that's a total of 1,367 GB of data.
- At 15.6 pounds per gigabyte, that's more than 10 tons of CO_2e.
- Mobile data, with its reliance on 3G/4G, is up to five times more polluting—77 pounds CO_2 per gigabyte.[7]
- If one million mobile users on 3G download a 1.4 MB page, that's 1,367 GB times 77 pounds, which totals 52 tons of CO_2.

---

4 James Christie, "Sustainable Web Design", A List Apart, September 24, 2013. (http://alistapart.com/article/sustainable-web-design)

5 Cody Taylor and Jonathan Koomey, "Estimating Energy Use and Greenhouse Gas Emissions of Internet Advertising", February 14, 2008. (http://evanmills.lbl.gov/commentary/docs/carbonemissions.pdf)

6 US Environmental Protection Agency, "Energy and the Environment". (http://www.epa.gov/cleanenergy/energy-resources/calculator.html)

7 Rainer Schoenen, Gurhan Bulu, Amir Mirtaheri, and Halim Yanikomeroglu, "Green Communications by Demand Shaping and User-in-the-Loop Tariff-Based Control", Proceedings of the 2011 IEEE Online Green Communications Conference (IEEE GreenCom'11). (http://ieeexplore.ieee.org/xpl/login.jsp?tp=&arnumber=6082509&url=http%3A%2F%2Fieeexplore.ieee.org%2Fxpls%2Fabs_all.jsp%3Farnumber%3D6082509)

In a subsequent interview, James notes that better figures must exist somewhere. In a perfect world, he says, it would be a simple formula, something like:

> Kilowatt hours per megabyte of data sent multiplied by the carbon intensity of the prevailing source that provided the power.

That simple formula, though easy to understand, doesn't take the amount of electricity burned by the end user into consideration. Nor does it take into consideration the amount of energy needed to upload and store the data on servers. Digital products and services have many moving parts and the stats James mentions in his article are constantly changing. For example, the 1.4 MB average page size he quoted in his post is, as of this writing, over 2.4 MB, according to the HTTP Archive.[8]

James notes that the moving target of defining the environmental impact of any online product or service is dependent on three things:

Storage and transmission
How much energy is needed to power servers and data centers, the telecommunications infrastructure that manages transmission, and the delivery mechanism (for example, 3G uses much more energy than hardwired Internet)?

Electricity transmission
How much energy is lost in transmission as well as how much energy is used by things like cell towers and electrical transformers to get data from its point of origin to the end user?

End-point use
What is the amount of energy needed to get data the final meter to the home or device, including how much power that device uses and if there are increases/decreases in power use depending on content served (Flash, for instance, can make people's machines work much harder and hence use more power)?

---

8 HTTP Archive, "Interesting Stats". (*http://httparchive.org/interesting.php*)

FIGURE 7-3.
Figuring out emissions produced to bring the game to your many devices is no simple task

Similarly, here's how sustainability consultant JD Capuano suggested proposing a website sustainability framework to estimate an organization's digital footprint:

> As James himself states, estimating a site's carbon footprint is tricky and imprecise even on the best of days. The long and short of it is I'd get more recent data and include alternative sources of data to tweak and add calculations to adjust for more realistic usage.
>
> First, are we just talking about the United States? If we are, data released by the government is always dated because it takes time to collect and update. If I were adding rigor to the process, I'd try to track down more recent figures. Similarly, we are in an era when coal plants are being shut down and natural gas plants are coming online to replace that baseload power. Renewables are also coming online each month, though still a small percent of overall power. I would look for related trends that are quantifiable and assess what adjustments could be made, and if none, at least footnote the known differences. I'd also keep an eye at fugitive emissions from natural gas extraction and

distribution, which may affect overall emissions reporting in the United States. If we aren't talking just about the United States, things get a lot more complicated.

Second, I'd revisit the mobile figures. Mobile plans have cut back our data and we don't want our data throttled back, so people I know use WiFi whenever they can while on mobile. Instead of just looking at 3G/4G network usage, I'd adjust for the percent of time mobile users are on WiFi instead of the network. This would still be imperfect, but better than a blanket approach.

Third, I'd ask if we want to measure the carbon footprint for viewing an average webpage, or the carbon footprint of the average webpage view. These are two different things. If we look at the latter, I'd want to break out sites with the most views—social media sites, Google's digital ecosystem, Amazon, etc.—and which of those have renewably powered (or free cooled) data centers. This would allow for an adjustment of Christie's input of emissions related to transmitting 1 GB of data for the term of the equation related to data centers. I would be curious to see if this would result in a meaningful reduction, as in something bigger than a really small decimal. I think you may be more interested in an organization's website and not all web pages (email, social media, etc.), but I still find this thought fascinating. Measuring both would help determine how well we're collectively moving the needle.

Building off that last point, if there are industry standard metrics such as I suggested, introducing elements like virtualization, caching, etc., and getting actual measurements from sites would be the way I'd like to see this proceed. We're a long way off from that and would need cooperation of cloud hosts, etc., but I think it's the direction in which we need to move.

Shawn Mills from Green House Data states that rough estimation is probably the best we can get:

> Like any calculation of this sort, it wouldn't be completely accurate, as you'd have to rely in most cases on average energy consumption of a personal computer or a percentage of overall energy consumed by a data center over the period of time the app is in use, those types of statistics. If you knew everything about the company and their equipment, plus the users' equipment, you could get more accurate.

You have to take into account the development time (computers used to create the application, the network traffic consumed, data centers resources, local backup—all of these have energy consumption). Then the energy used by the application itself, which is the data center load plus the users' own devices, plus the network traffic between them. I've seen wildly varying studies on the average energy consumption of one gigabyte of network traffic. So this would be a difficult metric and you'd have to make some assumptions. Even on just the network traffic, you have WiFi versus wireless broadcast versus different types of wired connections…it gets complicated very quickly.

The easiest way to get a ballpark, which is what most carbon footprint–type calculations are providing anyway, would be recording the time used on the app or website, the average energy used per gigabyte of network traffic depending on the connection type, plus the average consumption of the type of device being used (PC, laptop, smartphone…), plus the percentage of the data center's overall energy use attributable to the provisioned resources used for the application. Then depending on the energy source used to power all of those components you could adjust the overall footprint.

FIGURE 7-4.
Calculating the environmental impact of our digital products and services is no easy feat, but worth the effort once we figure it out

John Haugen from Third Partners, a New York–based sustainability firm offers the following approach:

> Like any carbon footprint assessment, the first step is to break down the process into assets and actions. For this digital example, the assets are all the digital data and physical equipment that are needed to produce the page. The actions are all the steps that must be taken to transmit data from the server to the user's device. Each action requires electricity, and the assets involved determine how much electricity is needed. The overall carbon footprint is a function of:
>
> - How many times those assets were called
> - From what fuel source is the electricity generated
>
> All carbon footprint assessment studies have scope boundaries, primarily to exclude both immaterial aspects and difficult-to-measure assets or actions. In many cases—and especially with complex digital scenarios—it is necessary to use proxies, estimates, and previous research. The reasons for this are both a matter of technical and economic limitations.
>
> Any website carbon footprint should consider pageviews, page size, device type of end user, and any content hosted outside the website by a third party (e.g., YouTube, Soundcloud, Dropbox).
>
> The approach to a mobile app would be similar, except it's broken down into two action categories.
>
> - The initial application download and all subsequent application update downloads
> - Use of the application
>
> Since the device stores a portion of the application content locally, the digital footprint of calling on local application data is limited to the usage of the device. Of course, any external data would still rely on the Internet for transmission to the device. And since mobile devices and tablets use per-minute a fraction of the electricity of personal computers, the degree to which the app can be operated locally without additional data will have a direct impact on its usage carbon footprint.

Product Science is a London-based agency that partners with organizations that are working on social or environmental problems as part of its business. Here is what its founder Chris Adams has to say on estimating a digital product or service's footprint:

> I would split it between the main phases:
>
> - Initial travel and work on the project—commuting to an office, and sitting in air-conditioned rooms uses noticeable amounts of energy.
> - Once the site is up, I'd measure how much goes into keeping it online, including any external monitoring or analytics services.
> - I'd also look at the cost of transmission—how much traffic is going to mobile versus desktop, and through that, how much is likely to be traveling wirelessly over 3/4G networks, as opposed to homes and offices.
>
> For a mobile app, I'd consider the same factors, with the adjustments for increased cellular network use.
>
> If you were to try to do this really thoroughly, and factor in the full life cycle of the devices involved, you would amortize the cost of the servers used to run a site, and the typical hardware upgrade cycle of the devices you're using. This gets really hard, and I'm not aware of any published examples of this.

Finally, I noted at the beginning of the chapter that someone owned the US patent on a digital carbon footprint calculator. After initially filing in 2007, in 2014, Alex Wissner-Gross and Timothy Michael Sullivan received US patent number 8,862,721 B2 for an environmental footprint monitor for computer networks. The methodology outlined in the patent includes the following steps:

1. Viewing a website embedding with a unique identifying tag from a user terminal.
2. Identifying the tag associated with the website.
3. Placing a cookie on a user's terminal.
4. Updating the cookie on the user's terminal with time data.
5. Updating the environmental footprint detection database.
6. Calculating the environmental footprint of the website.

This is, ostensibly, the methodology the owners used to create their website footprint calculation tool, CO2stats.com, discussed later in this chapter. The baseline for this would be the time at which the aforementioned unique identifying tag is placed on the website, which means that although you would receive important data on your website's footprint based on usage, you would still need to assess cradle-to-cradle considerations if you wanted to ascertain a comprehensive impact estimate of its entire virtual life cycle. Plus, given that the patent application was originally submitted in 2007, one has to wonder how the energy footprint of mobile devices factored into this methodology.

All this said, my original intent for this chapter was to find a solid formula for devising an accurate estimation of a digital product or service's carbon footprint. After all my research and given the complexity of this undertaking as well as the differing opinions on how to approach it, my best answer to a question of how to estimate the carbon footprint of a website would unfortunately have to be the answer that consultants and lawyers give (that everyone hates)....*it depends*.

Ecograder: A Case Study

As I mentioned in the book's Preface, my company, Mightybytes, really began paying attention to Internet sustainability issues around 2011, after becoming a certified B Corp. At that time, we explored different ways in which to use our own skills and talent as a force for good. Given the services we offer to clients, focusing on the Internet's environmental impact made most sense. Plus, I read Greenpeace's annual *Clicking Clean* report for the first time in 2012 as well as Pete Markiewicz's article "Save the Planet Through Sustainable Web Design," which really influenced our thinking about these issues.

Thus, we began figuring out how to apply what we learned during the process of becoming a B Corp to our *entire* business, not just the environmental impact of our office supplies, lighting, and so on. We started offering better healthcare benefits, paid volunteer time, and many other things companies do as part of the B Corp certification process.

We were also just getting underway on what would eventually become a long, drawn-out saga trying to find reliable web hosting powered by renewable energy, so the Internet's environmental impact was front-of-mind at the time.

During the 2012 B Corp Champions Retreat, the annual conference for B Corps that took place that year in Half Moon Bay, California, attendees were given downtime to connect and brainstorm ideas for using business as a force for good. During an oceanside stroll with a coworker and fellow B Corp owner, the idea for Ecograder was hatched.

FIGURE 7-5.
The B Corp Champions Retreat, a great place to hatch new ideas for using business as a force for good

VISION AND GOALS

Our hope was to create something as easy to use as HubSpot's Website Grader—put in a URL and go—that would help website owners, designers, and developers better understand that, however small, their site has an environmental impact.

We also wanted Ecograder reports to provide easy-to-understand, actionable roadmaps for improving efficiency in website performance, "findability," and user experience (UX), given that they were natural extensions of a more sustainable website. Of course, not everyone understands CSS sprites and content delivery networks (CDNs), but site owners could easily share the report with web teams using a simple link displayed directly below the score.

Another long-term goal was to connect users to green web hosts or resources for carbon offsets. At the very least, we wanted users to better understand why powering their site with renewable energy is important.

Finally, while the minimum viable product (MVP) for Ecograder was based on the idea of building awareness, the long-term business case should also be sustainable.

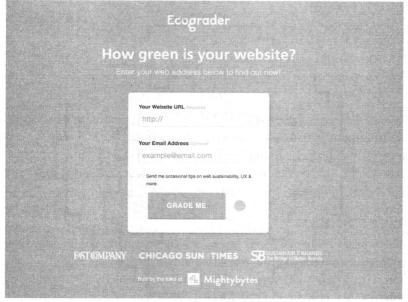

FIGURE 7-6.
Inspired by HubSpot's Marketing Grader, we wanted Ecograder to be a simple process: input your URL, click a button, get a helpful report

THE BUSINESS CASE

Because Ecograder was an app built by an agency, its business case was a little different than your standard startup. Even though we had aspirations of eventually making Ecograder a freemium model (free for basic services, pay for others), there were other important considerations, as well:

Awareness
> Ecograder should first and foremost bring awareness regarding the Internet's environmental impact.

Profit
> It should have a roadmap for income generation:
> - As a lead generator for Mightybytes.
> - Potentially as a Software-as-a-Service (SaaS) product.

Education
> The process of building it should educate our team on Agile and Lean startup methods.

Advocacy
> The launch, which we decided to do on Earth Day, should encourage companies to think more sustainably about their online properties.

To date, we have made good on all the preceding considerations except for finding the elusive SaaS revenue stream. In talking to a wide variety of sustainability consultants, general consensus was that for numerous reasons, lack of awareness among them, Ecograder's value proposition isn't something people would pay for.

THE METHODOLOGY

The biggest challenge by far was to devise a product methodology that increased awareness while providing actionable reports and also staying within our tight timeline, budget, and resource constraints. In other words, how could Ecograder quickly crawl a website without losing users and still provide easy-to-understand data that can help make sites more sustainable?

We knew with our constrained resources and timeline it would be impossible for the product to crawl an entire website and estimate the entire carbon footprint. For all the reasons mentioned earlier in this chapter, we decided that estimating a website's actual carbon footprint would not be within the scope of our MVP. To do so would undermine the "input your URL and go" simplicity we wanted and we simply didn't have the resources to take on a project of that scope.

FIGURE 7-7.
The methodology we used for Ecograder's algorithm followed the same process we used for building more sustainable websites

We also knew that many of Ecograder's desired metrics could be generated by existing tools with APIs we could tap into and extract relevant data. Here's the methodology we came up with:

Ecograder analyzes the contents of a web page (HTML, CSS, JavaScript, images, and hosting information) and runs a series of tests to compile a score. We devised several tests to help determine your cumulative score. These tests are meant to answer the following questions:

- Is your site hosted with a hosting provider that powers its servers by renewable energy?
- How many HTTP requests are made?
- Is your page optimized using industry standard methods?
- How easy is it to find your website?
- Will your site work on mobile devices?
- Did you avoid using Flash on your site?

Each of these tests produces a single score that is then weighted to help produce the final output score of Ecograder.

GREEN HOSTING

Ecograder scores green hosting high in its algorithm because the single biggest positive effect your website can have on the environment is to host it with a provider that runs on 100% renewable energy. After all, the servers that host websites require power 24 hours per day. Our challenge was to decipher whether a site was hosted by a green host and then score accordingly.

2.23 % of sites powered by renewable energy

FIGURE 7-8.
After three years of crawling websites, less than 3% were powered by renewable energy

Green hosting methodology

We initially created a homespun list of hosting providers that powered their data centers with 100% renewable energy. That quickly proved problematic (more on that in a bit). We then tapped into the Green Web Foundation's API to access its much more comprehensive database of green hosting providers. Through data provided by its API, Ecograder gives 15 points for providers that used renewable energy credits (RECs) or offsets, and the full 25 points for onsite renewable power.

Performance optimization

The challenge was to score sites based on how quickly they loaded and whether they followed standard practices for performance optimization. We decided that Ecograder would measure performance optimization in three ways: running a site's Google PageSpeed Insights score, measuring the number of HTTP requests the crawled page makes, and assessing the site's use of shared resources.

FIGURE 7-9.
Google PageSpeed Insights provided most of the necessary data to assess whether or not a site is optimized for performance; here are the averages of nearly 70,000 sites

Google PageSpeed insights

This easy-to-use tool from Google offers recommendations for improving the performance of any web page. It also provides an overall score based on how many best practices are followed by a crawled page. We decided to use this overall score as a metric in Ecograder's algorithm.

HTTP requests

To render a page in a browser, HTML pages send requests via HTTP to the hosting server for page components like stylesheets, images, JavaScript, and so on. Each HTTP request takes a certain amount of energy to complete and time to load, so more requests means longer wait times for content and more energy wasted. Reducing HTTP requests can improve both site performance and energy efficiency. Ecograder counts HTTP requests and assigns a score based on averages.

Shared resources

When you surf the Web, many of the sites you visit pull resources from common frameworks. Enabling your browser to access any resources that are already cached—rather than downloading them again—saves time, energy, and bandwidth. Ecograder assesses use of shared resources and assigns a score based on percentage of the page's total weight that is shared.

Findability

Our challenge was to help Ecograder determine how easy a site's content is to find on the Internet. Sites that are optimized for search help users find what they need faster because their content ranks higher in search engines. When users input search terms, they shouldn't have to navigate dozens of pages to find what they need. The process uses less energy when sites are search-optimized. We decided on MozRank to assess site findability.

FIGURE 7-10.
A site's MozRank provides data to Ecograder on how well your site ranks in search engines. Most sites crawled by Ecograder since 2013 scored pretty low.

MozRank

Created by Moz, a Seattle-based inbound marketing software company, MozRank is a "general, logarithmically scaled 10-point measure of global link authority (popularity). Because measures like MozRank are global and static, this ranking power applies to a broad range of search queries rather than pages optimized specifically for a particular keyword." Like Google's PageRank, MozRank takes advantage of the democratic nature of the Web. Your rank reflects the importance of your web page on the Internet. Ecograder uses your page's MozRank to add points in this category to its scoring algorithm.

DESIGN AND UX

Design is already subjective in nature, so the challenge of using software to assess a site's usability proved quite a challenge when devising scoring metrics for this Ecograder category. We explored many options, but we landed on two: mobile optimization, and whether a site uses Flash. The latter is less relevant now than it was in 2013 when we built Ecograder, but plenty of video and ad-heavy sites still use Flash to embed that content, even though in 2014 Google began penalizing sites that do this.

Mobile optimization

For the purposes of scoring, we define mobile optimization in two ways: sites that have separate experiences for mobile devices, and sites that are built with responsive design. Both of these approaches result in more energy-efficient sites because they both require streamlined content and design assets that load quickly. Ecograder analyzes the site's contents and qualifies a site as being mobile optimized if any of the following criteria are met:

- The stylesheet declaration contains the term `min-width` or `max-width`
- The stylesheet contents reference `@media`
- The site returns different *style.css* files when the user agent string is set to emulate a mobile device.

FIGURE 7-11.
Mobile optimization and Flash detection were used to discern whether or not a site provides a good experience for users across devices

Use of Flash

As noted earlier in this section and in Chapter 5, Flash consumes more energy than standards-based equivalents and doesn't work on most mobile devices, including all Apple and Android operating systems. From both an energy efficiency and performance standpoint, Flash is not the way to go, so we programmed Ecograder to look for embedded *.swf* files on pages it crawls and penalize a site's overall score accordingly if it finds any.

THE PROCESS

Ecograder went from concept to MVP in 11 weeks. Though the initial product was far from perfect, we considered it a success for many reasons. For the first time, with a few fits and starts, we used Lean/Agile methods on a project start to finish. We attempted to validate every hypothesis along the way through sprints while also sticking to an aggressive production timeline. We collected many email addresses. We even got a few client leads.

FIGURE 7-12.
Concept to MVP in just 11 weeks

Here are some things we learned in the process:

- The 11-week timeline was aggressive—it caused some stress, but also motivated our team to rally around getting an MVP to market quickly without sweating every little detail.
- Folding UX deliverables into product sprints driven by features proved challenging for our design team at the time.
- Working with reporter timelines while also sticking to an Earth Day launch made for some long days.

It is also important to note that we were juggling client work at the time, too. Most work on Ecograder happened in dedicated work sessions purposefully scheduled around client deadlines.

Over the contracted timeline, we focused on research, competitive analysis, sprints, and design and UX. The following sections look at each of these areas more closely.

Research

For starters, we interviewed dozens of agencies, sustainability practitioners, and B Corp community members to ascertain whether the idea was viable and if we could find a product-market fit. Feedback was encouraging, though many expressed trepidation over the value proposition being something for which they would consider paying.

Most people didn't realize that Internet sustainability was a thing they should pay attention to and noted that the interview process opened their eyes. Based on the awareness factor, target users encouraged us to proceed.

Competitive analysis

We ran a competitive analysis and found others trying to address these same issues:

Greenalytics[9]

>An open source tool developed at KTH Royal Institute of Technology in Stockholm, Sweden, Greenalytics mashes up data from Google Analytics and environmental research to estimate the carbon footprint of websites, including server, infrastructure, and final user. Its data is open source and posted to GitHub. The last GitHub update was made in 2013.

CO2 Stats

>A pay-per-month service, CO2 Stats calculates your website's environmental footprint, finds ways to make your site more energy-efficient, automatically neutralizes its calculated carbon footprint, and displays a green-certified "trustmark" on your site.[10] Owners of this site own the previously mentioned patent on calculating a website's carbon footprint.

Web Energy Archive Ranker[11]

>Created by Green Code Lab in France, this tool pulls data from the HTTP Archive, Webpagetest, and the Power API to track Internet energy use trends and raise awareness of energy consumption and environmental footprint. Of the three resources noted here, this one seems to still be an ongoing concern and provides the most comprehensive information.

These tools were critical in helping us stay focused on simplicity in usability.

---

9   http://greenalytics.org
10 http://www.co2stats.com
11 http://webenergyarchive.com/en

FIGURE 7-13.
Other Internet sustainability tools

7. DIGITAL CARBON FOOTPRINTS | 253

Sprints

Sprints were driven by specific integrations and occurred in one- to two-week timeframes, depending on our existing client workload and complexity of the integration. After we had an initial roadmap based on the methodology just outlined, we were able to quickly pull data from the APIs we identified.

Design and UX

As I noted earlier, our design team at the time struggled with how to incorporate deliverables like wireframes and visual comps into the Agile processes we used to build Ecograder. They also struggled initially with making more sustainable design choices. As Pete Markiewicz notes, "Designers can't just follow their muse—they must balance against the needs of their audience (UX) and the environment (design carbon footprint)."

We've since learned that challenges integrating design practices is common for teams just getting started with Agile methods and have incorporated techniques such as those outlined in Chapter 5 to address this. Of course, we have become better at incorporating more sustainable choices into our design process overall, as well.

CREATING ECOGRADER CONTENT

To make Ecograder's reports more useful, we wanted to link out to additional resources that could help site owners and product users make their sites more efficient and environmentally friendly. But the problem was that most of those blog posts didn't exist anywhere on the Internet. Sure, there were posts all over about how to use techniques like CSS sprites, shared libraries, and so on, but none specifically addressed why those were more sustainable strategies. Our content team went to work creating a series of posts to bridge the gap between Ecograder's metrics and efficiency-driven web design techniques.

PROMOTING ECOGRADER

It was mentioned previously that advocacy was part of our strategy for Ecograder. Not only did we want to use these techniques on our own sites, but we wanted others to embrace them too…or at the very least, be aware of them. To accomplish this, our content team worked with a PR team to run every one of the Fortune 500 companies' websites through Ecograder and track the scores. We shared this data with a few members of the press, which resulted in an article on Fast Company's Co.Exist blog titled "Measuring the Efficiency of Fortune 500 Websites." It was also covered by a number of other blogs and media outlets, including *Chicago Sun-Times* and Sustainable Brands.

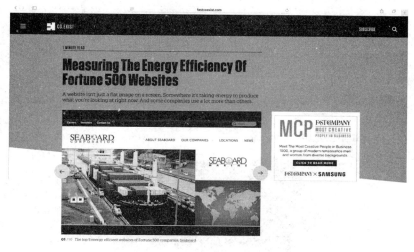

FIGURE 7-14.
Running the Fortune 500's websites through Ecograder

RESULTS

Although we haven't yet cracked the nut of how to make money with Ecograder, it has generated positive business results for Mightybytes. Above and beyond the aforementioned press coverage, it has provided other benefits:

- We have used it to benchmark site performance for clients who care about sustainability.
- It has crawled thousands of URLs and connected us with thousands of users all over the world who are using it to benchmark the creation of more sustainable websites.
- It gets regular mentions on social networks.
- For much of the first year after launch, it was one of the highest ranking pages on our site.
- It has resulted in speaking engagements around the United States, a TedX Talk, and was part of the pitch for writing this book.

On Earth Day 2016, three years after its initial launch, we collected data from nearly 70,000 website crawls by Ecograder and created infographics with the data, including that slightly more than 2% of the sites crawled were hosted on servers powered by renewable energy and only 24% were optimized for mobile devices.

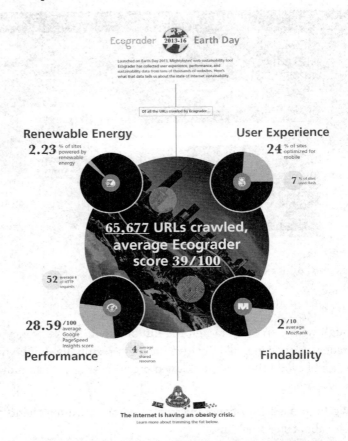

FIGURE 7-15.
Compiling data from three years of crawling websites for sustainability

For all these reasons, we consider Ecograder a great success. It met nearly all of the considerations in our business case. It has helped differentiate Mightybytes from the hundreds of thousands of other agencies out there that do what we do, and it has raised awareness on the issue of Internet sustainability.

In keeping the entire product life cycle in mind, was the process of creating Ecograder as sustainable as the product itself? For the short term, I'm sure there were many things we could have done more efficiently. We could have been more Lean in our user research. We have hosted it on a handful of different green web hosts, with mixed results. The design has evolved over time. Over the long haul, however, creating Ecograder helped our team become more proficient at Agile methods and we gained consensus on which design and development techniques should be considered more sustainable than others. All in all, the learning opportunities far outweigh any extra resources we might have used during its creation.

Ecograder Benchmarking

Here's how we use Ecograder to help clients improve their websites. Climate Ride is a US-based nonprofit that produces charity endurance events that raise money for the environment, active transportation, and sustainability. Participants register for these transformative, multiday events and use peer-to-peer fundraising techniques to raise money for one or more nonprofit beneficiaries. Like many small nonprofits, the organization takes a Lean and iterative approach to its growth while always paying attention to people and planet as they scale. Its website is no exception.

Mightybytes redesigned Climate Ride's original website back in 2011. At that time, as part of the redesign, the site was migrated to a hosting provider powered by 100% renewable energy. Two years later, we ran the site through the first version of Ecograder. Even though Ecograder didn't provide every metric needed to make Climate Ride's site more efficient, its scoring mechanism was enough to help its team better understand how recommended improvements would make its site more efficient and environmentally friendly while also enhancing UX.

FIGURE 7-16.
Thousands of riders have raised millions of dollars for their favorite environmental nonprofits through Climate Ride

Climate Ride's initial Ecograder score in 2013 was 71. Because Ecograder gives 25 points out of the total 100 available for green hosting, we can safely say that their score would have been at 46 prior to 2011. This low score was not at all a reflection of the organization's overall environmental impact; however, it was a great place to begin benchmarking website sustainability.

FIGURE 7-17.
Over time, Mightybytes made many under-the-hood tweaks that resulted in a faster, more efficient site

In 2014, we again updated their website to continue improvements to the sustainable web design concepts outlined in this book. The following sections outlines some of the things we improved.

FINDABILITY/SEO

Mightybytes streamlined the content structure of Climate Ride's site to make popular content types easier to find, cutting down on user search time. We continue to improve the site's search engine results by optimizing pages, helping them create qualified inbound links, and so on. Also, the scenic beauty of Climate Ride's events makes photos a big part of its online marketing strategy, so Climate Ride performs very well on social media, which is the second highest referral channel behind direct traffic. As soon as users from social networks—like all users—are on the site, our efforts help them get to the content they need quickly.

DESIGN AND UX

Similarly, common user interactions such as finding a rider, registering for an event, or donating were given more prominent placement in the site interface, again in an effort to get users to the content they need quickly. Responsive design improvements helped make the site more mobile friendly.

PERFORMANCE OPTIMIZATION

Large background images, which each triggered a separate server request and ranged anywhere from 60 to 300 KB per image, were removed.

The home page slideshow, which sometimes contained up to 10 images, was replaced with a single, lightweight hero image, significantly cutting down load times and server requests. The Mightybytes team also introduced Climate Ride to tools like Smush.it, PicMonkey, and ImageOptim to further compress site images.

During this time, Mightybytes also updated Ecograder's scoring algorithm, adding more metrics to check against. After the site overhaul, Climate Ride's website scored a 91, a notable improvement of 45 points over the 2011 website.

FIGURE 7-18.
Benchmark this: over several years, Climate Ride improved its website's Ecograder score by 45 points

As a small, virtual nonprofit working primarily with online tools, Climate Ride has always focused on implementing sustainability in all aspects of its operations. It is committed to providing optimized, energy-efficient online communications that serve our community's needs and, whenever possible, are powered by 100% renewable energy.

Conclusion

We covered just how complicated it is to estimate a website's carbon footprint in this chapter. Herein lies a big challenge with Internet sustainability. With the Greenhouse Gas Protocol, we have a standard process for estimating greenhouse gas emissions. We could apply that process to a business or even to a data center and generate a relatively accurate estimate of that entity's emissions. This would give us a better understanding of that entity's environmental impact in terms we could easily understand.

But if you're talking about an end-to-end assessment that includes all the potential variables where a digital product or service might have potential environmental impact—from the process of its creation to its hosting and distribution through the infrastructure that serves it to end-user devices—this becomes an incredibly complicated undertaking. Until someone cracks this nut, however, it is difficult to talk about Internet sustainability in terms we can gain consensus on and easily digest.

In the meantime, if someone wants to fund a research grant, this would be a good, meaty project.

Action Items

Try these things:

- Use Ecograder to benchmark your site score.
- Identify a list of potential improvements.
- Improve your site and track score improvement.
- How would you estimate your website's carbon footprint?

[8]

A Future-Friendly Internet

What You Will Learn in This Chapter

If all digital products and services were efficient and low impact, if all hosting was powered by renewable energy, how would a more sustainable Web work for years to come? This brief chapter closes the book on a bright note, offering a glimpse into what a sustainable Internet might look like down the road.

FIGURE 8-1.
Will the digital future be green?

Future-Friendly Web

We have spent many pages in this book talking about how to make our digital products and services more lean, future-friendly, accessible, and sustainable. If we're going to make the reality of an efficient Internet powered by renewable energy and run on devices created with fair and equitable practices, some big things need to happen. The following sections discuss some ideas for how we can get there based on topics already covered in this book.

CONSCIENTIOUS COMPANIES

Online commerce plays an increasingly larger role in business being the most powerful man-made force on the planet. Thus, the future of the Internet is intricately woven into the future of business. Thankfully, there are some rapidly growing movements afoot with the potential to transform business.

Groups such as Conscious Capitalism and the B Corp movement are doing wonders to infuse business with a sense of purpose beyond just profit. Inspiring certified B Corps like Etsy, Kickstarter, Hootsuite, and other companies that earn their primary income from online sources are using the Internet to do well while at the same time solving social and environmental problems. Tools like the B Impact Assessment, which is used to certify businesses that want to become B Corps, go a long way toward helping companies embrace accountability, transparency, social responsibility, and environmental stewardship. Companies that embrace these values are far more likely to have more environmentally friendly company cultures, gender pay equity, open-book accounting, better benefits, paid volunteer time, and so on. In fact, compared to other sustainable businesses, certified B Corps are:[1]

- 68% more likely to donate at least 10% of their revenues to charity.
- 47% more like to use onsite renewable energy.
- 18% more likely to use suppliers from low-income communities.
- 55% more likely to cover at least some health insurance costs for employees.

---

1   B Lab, "B Corp Community". (*http://www.bcorporation.net/b-corp-community*)

- 45% more likely to give bonuses to nonexecutive members.
- 28% more likely to have women and minorities in management.
- Four times more likely to give paid professional development opportunities.
- Two and a half times more likely to give employees at least 20 hours per year paid time off to volunteer in their community.

Becoming a B Corp or working at a B Corp is just one way to work toward a more sustainable planet for people and pixels. Beyond B Corps, businesses that simply want to offset their emissions can work with companies like ClimateCare, another B Corp that helps public and private sector partners improve their environmental impact. You buy offsets or renewable energy credits (RECs) to cover your emissions, they deliver programs designed to tackle poverty, improve health, and protect the environment.

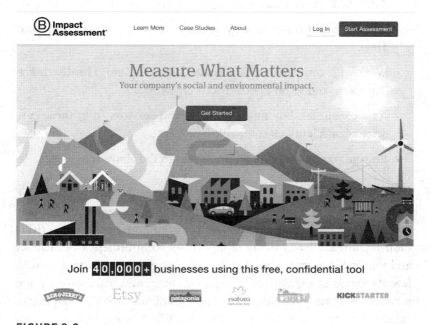

FIGURE 8-2.
Measure what matters online and in your business

The number of these companies around the world is growing, and although only a segment of them currently consider digital as part of their environmental assessments, these companies are hardwired to consider impact and innovation in tandem with existing business practices. With the right standards and solutions in place, it is only a matter of time before B Corps and other conscientous companies consider digital as a core part of their environmental impact.

It should also be noted that business isn't the only sector addressing these issues. Nonprofits like The Green Grid, the Green Web Foundation, the World Wide Web Foundation, UNESCO, BSR, and many others are working diligently to bring accessibility, equity, efficiency, and renewable energy to the Internet.

To build a more sustainable future, we need these sectors working together for collective impact. A more people- and planet-friendly Internet is part of the equation.

Here are some small things you can do right now to be part of the growing movement of people using business as a force for good:

- Take the B Impact Assessment (*http://bimpactassessment.net*).
- Join the local Conscious Capitalism chapter (if there is one in your area).
- Work with companies like ClimateCare, TripZero, ThirdPartners, 3Degrees, and so on to offset your emissions.
- As a consumer, buy more products and services from these conscious/conscientious companies.
- Engage your co-workers or employees in the above activities.

GREEN HOSTING

Green hosting is one area where significant progress can be made quickly. The information and communications technology (ICT) industry already leads many others in its commitment to efficiency and renewable energy. However, it is also a very opaque industry with few standards or regulations. For public companies, transparency and accountability are often sidestepped in favor of maximizing shareholder value (in the United States, that's the law). Whereas Apple, Google, and Facebook have led the way in powering their data centers with renewables, hundreds of other companies still have a long way to go.

To date, no standards or comprehensive systems exist to help potential customers easily understand whether a particular data center is green or if it is just greenwashing. Some companies, like Green House Data and Canvas Host, go to extensive measures to help consumers understand their commitment to efficiency and powering their servers with renewable energy. For many internet service providers (ISPs), however, it is either not a priority or they have insufficient resources to measure or improve their energy performance—or they simply choose not to disclose their consumption.

To do its part in building a greener Internet, the web hosting industry should consider creating standards similar to power usage effectiveness (PUE) or data center infrastructure efficiency (DCIE) for powering data centers with renewable energy. Purchasing RECs is simply not enough. Web hosts (and all companies for that matter) should use their influence to affect legislative change at the local level and bring more renewables into their region so that they can power their servers *directly* with renewable energy. Again, efficiency and renewable energy must be the key core components of more sustainable strategies.

David Anderson of Canvas Host notes: "As there are no clear set standards, it is difficult to suggest one set of proposed standards versus another. There are thankfully many new and increasing numbers of organizations like Bonneville Environmental Foundation, and I would recommend any service provider reach out to find if there are organizations like this in their own immediate vicinity, or at least state, as that can help identify what programs are in place or proposed, and ways that company can get onto their own sustainability path."

David says he feels the metrics established for his business are important insofar as they relate to immediate hosting statistics, but they don't address other operations components, such as recycling and e-cycling:

- What and how much are we recycling in terms of physical hardware or plastic?
- What are we upcycling?
- Do we offset our operations through mass transit?
- What about local tree planting (which Canvas Host does through Friends of Trees)?"

FIGURE 8-3.
Sweden's Green Web Foundation is pushing hard to raise awareness of green web hosts

Sweden's Green Web Foundation has yet to become the Internet standard for green hosts, but its efforts have definitely created more visibility for existing green web hosts. "Visibility on the Internet is the only currency that counts, and we try to make green hosts more visible," says cofounder René Post. He notes that they have received many requests from companies wanting to be included in their database of green web hosts but whose only achievement is the use of a data center with a low PUE. "While that is important," he says, "the remaining energy use might be provided by coal and nuclear, which could still be sufficient to power a small town. So we praise them for their efforts, and inquire at the same time when they will switch to renewable energy so they can be really green."

René also notes that his organization's efforts to help green hosts more clearly communicate their environmental impact face significant challenges, as well. "Let's be honest, there will never be a time when we can be sure we have covered 'the whole of the Internet.' It is huge, branching out in many countries and languages, and the growth is still unbelievable. So that is a challenge, and we do not want to claim that it is comprehensive."

For example, as of this writing, the organization has been unable to find a green host in China. Right now, that means there is no active green host in China that is writing about its use of renewable energy in English. "Language is very important," he says. "We have mainly looked at the Arabic, Russian, English, German, Dutch, French, Italian, Spanish, and Scandinavian languages, but our understanding of Burmese or Urdu (to name a few) is severely limited. So we might miss some developments there at this time."

Product Science's Chris Adams thinks the larger cloud providers already have a big leg up on smaller web hosts when it comes to green hosting. "I think you give so many competitive advantages away when you choose a small, if well meaning, green web host that 'just' focuses on giving you a server to run. You need to be exceptionally good at providing operations, to make up for that loss in responsiveness or agility."

Chris notes that transparency and increasingly complex supply chains as companies offer more services for lower prices will prove to be a big challenge for a cleaner, greener Internet moving forward. "Even if I try to use only Google's servers, which run cleaner than most, we end up using between 5 to 10 external services and tools on a typical project. It's very difficult to ascertain how green these services are."

Though it can be difficult to find a good green web host, here are some things you can do right now:

- Check out ServingGreen (*http://serving.green*) and The Green Web Foundation (*http://www.thegreenwebfoundation.org*) for updated information on green web hosting.

- Use the US Department of Energy's Buying Green Power map to discover whether renewable energy is available for the region in which your hosting provider is located.[2] If so, pressure it into converting its energy source to renewables.

---

2 US Department of Energy, "Can I Buy Green Power in My State?". (*http://apps3.eere. energy.gov/greenpower/buying/buying_power.shtml*)

HARDWARE

Because this is a book about design and development, we have talked a lot about software, but it's important to note the role hardware plays in creating a more sustainable Internet, as well. All those devices and servers require electricity to run, very little of which, as we have established, comes from renewable sources. Plus, supply chains for the devices we use on the frontend and those that reside in the data center are often marred by environmental disasters.

Near Inner Mongolia, China, for example, a lake exists that is filled with the toxic by-products of refining rare-earth metals.[3] This lake is only a few miles away from farmland and a city of 2.5 million people. The mines that pollute this lake produce about 70% of the world's rare-earth metals, used to manufacture consumer electronic goods.

This is just one example of similar sites all around the world. As Green America Content Strategist Bernard Yu says about the computer or device you might be reading this on right now: "If any of the gold or zinc in your computer came from Bolivia, it probably used child labor. If it came from the Congo, it probably used child labor *and* funded wars."

FIGURE 8-4.
Byproduct of our devices: a lake in China filled with the toxic derivative of refining rare-earth minerals

---

3 See "Building More Holistic View of Sustainable: Digital Project Planning at Green America", available on YouTube. (*https://youtu.be/IW_qRxcQIe8*)

To this end, Greenpeace is working hard on parts of the Internet supply chain that require hardware: data centers, wireless infrastructure, and the footprints of the devices themselves, going all the way back to the manufacturers. "Many of the device manufacturers are in areas powered significantly by coal," says Greenpeace's David Pomerantz. "IT companies pioneered the ability to take control of their energy supply, which has since taken off in many other industries. Large IT companies showed everyone that it's good business. Now we need to look at the *entire* manufacturing supply chain for servers, phones, laptops, tablets, and so on."

One company leading the way toward more sustainable hardware sourcing is Amsterdam's Fairphone. Originally started as a campaign within an organization called Waag Society on conflict minerals used in cell phones, this social enterprise has evolved into building a movement for fairer electronics. Now a certified B Corp, Fairphone focuses on sustainability as the reason it exists and so its activities across the entire life cycle, from mineral extraction to design to manufacturing to end-of-life, are all driven by making more sustainable choices. The organization is 100% independently financed (no donations or venture capital), which lets it preserve its social values.

FIGURE 8-5.
By sourcing its product materials sustainably and committing to open source, modular design, and addressing social issues, Fairphone aims to revolutionize the smartphone industry

Here's an example of how Fairphone aims to change the smartphone industry. It sources e-waste out of Ghana into Belgium, where it manufactures parts for its devices. For every cell phone it brings to market, the company has a goal to source materials from three end-of-life phones from places without the infrastructure to support e-waste recycling. Though it is less sustainable to ship old phones from Ghana to Belgium, this process helps address a specific issue by helping Ghanaians build the right infrastructure to recycle e-waste in their country.

The company is also committed to open source and is about to launch an open source version of its product (Android-based). The company considers open source very important for its growth strategy and helps it bring more sustainability to its devices.

By telling these stories, the company also helps consumers see value in its products and is able to use business to change the world.

EDUCATION, INCUBATION

Awareness will play a huge role in moving a cleaner, greener Internet forward. This can't happen without the support of key influencers: public speakers, educators, business incubators, professional organizations, and so on. If the future of the Internet is to be sustainable, students and new business owners will be on tomorrow's frontline making it so.

Education

Educating students about the role of sustainability in design is key as those students enter the workforce and begin creating products and services. Most sustainable design programs focus on architecture. Sustainability should be wrapped into the curriculum of all school programs rather than isolated within certain departments. Professional development organizations such as AIGA and GDC already tout the benefits of accessibility and sustainability. Why not take the next step and begin integrating sustainable design techniques with digital products and services?

Dr. Pete Markiewicz notes that students often have the skills or technical knowledge necessary to create digital products and services but lack context for their decisions. "A typical student working in web design or web development might follow a personal 'green' lifestyle," he says, "but create bloatware for the Internet. The reason is that the Web feels 'weightless'—it is light, airy, and doesn't seem to 'dirty' the environment in the obvious way an automobile can."

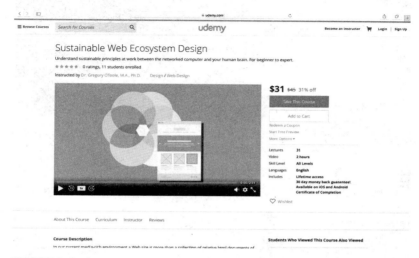

FIGURE 8-6.
Education on the environmental impact of the Internet will be critical to ensure its future is more sustainable

He also says that the role of educators in this area is to give students insights into two key areas:

Bringing more sustainable thinking into design
> "Design is the primary way and appropriate level to introduce sustainability thinking into products and services. Design choices have real consequences. Designers can't just follow their muse—they must balance against the needs of their audience (UX) and the environment (design carbon footprint)."

Personal responsibility
> "Design becomes unsustainable right at the level of each individual who directly creates products and services. If that's your job, you are personally responsible for implementing sustainable design practices—not your company, industry, or government."

FIGURE 8-7.
Digital business is thriving—helping startups better understand these issues will make for more sustainable digital products and services in the long run

Incubation

Similarly, as more startups find success and long-term growth online, entrepreneurs will need to make critical decisions for scaling their companies sustainably. Business incubators and startup accelerators sometimes focus on product but neglect addressing long-term viability or impact of business decisions on all stakeholders. In the mad dash for startup funding, these principles can become lost.

Noel Burkman, partner at Month16, a business incubator that specializes in helping entrepreneurs create "minimum viable businesses" with greater chances for success, says startups that don't focus on business viability alongside product viability are putting their futures at risk. Here's what Noel has to say about the importance of incorporating sustainability into the startup process:

> Sustainability impacts every facet of business. Brands are built on it. It drives consumer choice. And long-term growth depends on it. I've hired many a designer and developer, and ALWAYS, ALWAYS, ALWAYS choose the ones that understand efficiency from the ground up. Sure you can learn to code in a month or declare yourself a web designer because you've built your own website. But if you don't understand the fundamentals of algorithms, or basic UI/UX principles, I guarantee you that the digital experience you create will be expensive. You'll be chewing up CPU cycles, which can get expensive at scale; your security will get expensive when breached and your bandwidth and load times will create a slowness that ultimately drives customers away. New businesses looking for rapid growth often overlook the importance of establishing sustainability policies on which a lasting, scalable foundation is built. Any group providing leadership and guidance to early-stage entrepreneurs prioritizing rapid product development before developing sustainable practices is building a house of cards.

ANALYTICS AND ALL THE THINGS

The Internet of Things (IoT) will offer ample opportunities to better monitor our energy use and make more educated decisions driven by data. Rapidly growing startups like Chicago's Uptake, whose platform uses data collected from sensor-enabled equipment and machines to help the world's companies tackle difficult problems, offer potential to redefine entire industries and significantly reduce emissions.

One of the IoT's significant contributions to a more sustainable society might be in its ability to contribute to the circular economy. According to architect and designer William McDonough, "The circular economy is an economic system that is an innovation engine that puts the 're' back into 'resources.' It allows for continuous benefit to be provided to all generations by the reuse of things, of material, energy, water, [etc.]"[4] In other words, IoT devices could help inform decisions about how best to reuse a product's materials or convert it into something new and usable at the end of its life cycle. All the data that smart devices collect could be used to maximize utility.

A 2016 report from the World Economic Forum and the Ellen MacArthur Foundation titled "Intelligent Assets: Unlocking the Circular Economy Potential" explores potential ways in which a digitally enhanced, prosperous circular economy could look.[5] It offers "fertile ground for innovation that could enable the decoupling of value creation from finite resource consumption and lead to broad social benefits."

It suggests that these intelligent assets could change how we make, use, and reuse products by doing the following:

- Monitoring performance through data transmitted from product to manufacturer
- Redefining maintenance through performance contracts, predictive maintenance, and automatic updates
- Helping designers and developers make better decisions through data-driven product improvements
- Extending the "use cycle" through parts harvesting, reuse, recycling, and so on

The report also discusses how these assets will create a localized, smarter energy infrastructure by implementing the following:

---

4 Sustainable Brands, "6 Technology Trends That Are Reshaping The World We Live In". (http://events.sustainablebrands.com/sb16sd/updates/6-technology-trends-that-are-reshaping-our-world)

5 Sustainable Brands, "Intelligent Assets: Unlocking the Circular Economy Potential", February 8, 2016. (http://www.sustainablebrands.com/digital_learning/research_report/next_economy/intelligent_assets_unlocking_circular_economy_potentia)

- Transmitting energy data from local renewable energy plants to communities generating their own energy
- Monitoring energy performance
- Creating private and community energy stores that service commercial and social/public spaces
- Encouraging "pay-per-use" renewable energy

The report also explores other potential benefits as well, including these:

- How increased data intelligence in buildings, bridges, and roads will allow for predictive maintenance, efficient use of resources, energy efficiency, intelligent reuse of demolition waste, as well as communicating real-time health data for buildings, infrastructure, energy, and other assets.
- Increased efficiency in transportation and logistics through optimized delivery routes, fleet efficiency, avoiding waste, sorting recycled materials, and so on.
- Enabling healthy, resilient fish stocks and regenerative farming.
- Smarter cities with asset tracking, materials tracking, sustainable domestic water usages, localized smart energy grids, energy-saving streetlights, and so on.

Of course, all of these intelligent sensors around every corner will have vast data transmission and electricity needs. The potential this offers for reducing environmental impact and benefitting society, however, is enormous. Let's hope we can power them all with renewable energy and design them as efficiently as possible.

VIRTUAL REALITY

Similarly, as virtual reality (VR) continues to grow in popularity, it too will have a huge impact on society's data and electricity needs (it goes without saying that the concurrent rise in popularity of online streaming media presents an equal challenge).

VR can also help increase understanding of sustainability by immersing users in real-world environmental and sustainability scenarios. Few people have seen a coral reef degraded from ocean acidification, for instance, but VR can put them up-close and personal with precisely how our behavior affects the environment.

Agency Atticus Digital created a VR training package for one of Europe's biggest energy companies to help employees understand why sustainability is crucial for business, society, and planet.[6] Using Oculus Rift headsets, employees can learn about the impact of issues like excess waste, irresponsible supply chains, deforestation, and overconsumption.

Finally, VR UX will no doubt also transform the way we interact with digital products and services. A virtual shopping experience could replace a trip to the mall, for instance. A virtual training experience could reduce the emissions created by a cross-country trip to a training lab.

ONLINE LEGACY

With all this in mind, what do we do with the billions of web pages that the first two-and-a-half decades of the Internet produced? As we discussed in Chapter 4, businesses are already hesitant to retire their content. If the current average web page is the digital equivalent of a Humvee, do we really need a massive fleet of them taking up server space and burning pixels across the Internet? Is there a way to make the great Internet content graveyard a little more sustainable?

Taking a page from the aforementioned circular economy, I would encourage all organizations to regularly audit their content, cutting waste and using analytics to inform decisions about what should stay and what should go. If all companies did this at least once per year, the Internet of the future would be leaner and cleaner.

---

6 Matthew Yeomans, "Oculus Rift Isn't Just a Game; It Could Have Powerful Effect on Sustainability", *The Guaridan*, March 26, 2014. (http://www.theguardian.com/sustainable-business/oculus-rift-facebook-acquisition-sustainability-effect)

Interviews: Industry Leaders Predict

I asked the folks I interviewed throughout this book for their take on what the Internet of the future might look like when viewed through the lens of sustainability. Here's what they had to say.

GREG HEMMINGS, HEMMINGS HOUSE

> I see more streaming, I see more cloud-based apps, I see untethered access to creativity tools. This is going to be very taxing on the grid. If we as B Corps and triple-bottom-line businesses continue to pressure the establishment to do better, I see the GREEN cloud and the GREEN grid powering our creative story production and sharing needs.

EMILY LONIGRO BOYLAN, LIMERED STUDIO

> I'm optimistic. I think it's going to be simpler and lighter. And I think that to be simpler and lighter it's going to require a lot of really smart people from around the globe to work on solutions. Good thing we know each other, right? And we can't wait for anyone to decide what to do, we have to see an idea, grab onto it, and execute. Agile.

MIQUEL BALLESTER SALVA, FAIRPHONE

> Technology companies have an opportunity to lead the way toward a more sustainable digital future. The number of devices is increasing greatly; as is the huge amount of data we produce. Building more media-hungry devices makes this even more challenging. Sustainability is the reason Fairphone exists, it's not just a driver. We think sustainability first. If more technology companies adopted this approach, we could significantly reduce our industry's environmental impact and that of the Internet as a whole.

SHAWN MILLS, GREEN HOUSE DATA

> I'd say the end game is an Internet (and other IT services) powered entirely by green energy and run on equipment that uses as little of that energy as possible. It's all about taking the available technology and maximizing its use so we don't waste energy and on the other end of things, we don't waste time, like waiting for a web page to load.

FIGURE 8-8.
A more sustainable Internet of the future will be analytics-driven and iterative, just like other industries

DAVID POMERANTZ, GREENPEACE

> Our future Internet is 100% powered by renewables. How do we get there? Greenpeace is already leveraging customer pressure but to reach the end goal we need to address every layer in the value chain: consumers, brands, Internet-enabled services communicating to their hosts, data centers, wireless infrastructure, and the footprint of the devices themselves, which goes back to the manufacturers. How can we help them create products with as little environmental impact as possible? Finally, we need to communicate the importance of these efforts to policy makers. Without environmentally friendly legislation, there's only so much we can do.

PETE MARKIEWICZ, THE ART INSTITUTES

A sustainable Internet will be very analytics-driven. All design (and by this I mean visual design, even down to individual elements in a graphic) will have analytics on it, estimating the cost of delivery versus benefit to the end user. Everyone will use analytics as a driving feature of their design. Even in the early stages of concept design, designers and developers will consider sustainability. Finally, late-stage development efficiency issues will feed back onto the design, causing iterations of everything, even Art Direction.

This change will signal the end of "postmodern" theory in design, and the dawn of the "age of the user" as the dominant art theory (see Ellen Lupton's *Thinking with Type*).

And this Web will have a better appreciation of the power of just plain text and typography, as opposed to elaborate motion graphics, movie-style experiences, or virtual reality.

Another issue is the Internet of Things. Many proposals call for computers in every lightbulb. This will cause a rise in energy use, even if the devices themselves are low-power. The place where we will need more energy is the monster server farms needed to create the cloud computing necessary to process all that IoT data.

A big challenge coming up is Virtual Reality and Augmented Reality. These tools will often be web-based (see *https://mozvr.com* for examples) but have the potential to really be resource hogs. The amount of resources needed to create immersive 3D worlds is bound to be higher than the 2D web. And with 3D tools that allow anyone to create, instead of the tiny minority learning Maya or 3DSMax, we will have an explosion of 3D-ish websites with monster footprints.

JAMES CHRISTIE, MADPOW

A sustainable Internet is one that is responsible for displacing far more carbon than it produces. In other words, it is inevitable that building sites and physical Internet hardware have carbon costs, but so long as they are enabling new paradigms to replace old polluting methods, they will remain an overall force for climate good. I think the Smarter2020 report said we'd save 9 GT by 2020 by this paradigm shift. But it has also been predicted that the footprint of the Internet by 2020 will be 1.4 GT. So, that's a saving to loss ratio of 6.4:1. Obviously it would be better if the ratio was more like 10:1.

We need standards and building codes to ensure physical infrastructure is built using sustainable techniques; hosting and telecomms powered by green energy, related buildings meet LEED; end-point devices (phones, etc.) made as greenly as possible (See Fairphone, etc.).

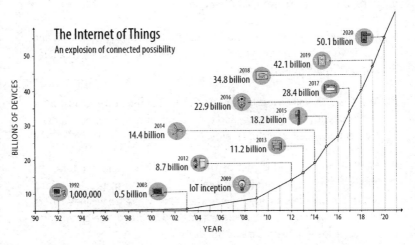

FIGURE 8-9.
Tweets on toast: How much unnecessary consumption will occur with an estimated 50 billion connected devices by 2020?

With the IoT, we're talking about billions of new devices connected to the Internet, with resulting energy needs to transmit the data, and perhaps a trend to keep devices "always on" so they can take advantage of IoT. In device terms, IoT brands are pushing for consumers to replace their perfectly good toaster with the latest Internet-enabled toaster so they can print the latest tweets onto their breakfast (this is a real thing!). This will result in more consumerist churn and unneeded consumption. Fortunately, there is also a backlash, as exemplified by the Internet of Shit Twitter account.

The best way to reach a more sustainable and future-friendly Internet is through environmentally friendly legislation, and shifting consumer standards. People demanding greener everything will drive change. And probably the need for a shift in consumption patterns towards less. That said, I'll leave you with these questions:

- How else can we use the Internet to foster international cooperation? We just signed the Paris COP21 treaty, which means each of the 196 countries has its own way to combat climate change. Can we do more to spread best practices?
- Can we build or influence online communities to be greener (beyond just putting badges on sites)?
- Can we arrive at a new normal where every ecommerce site labels the greenest products? Or where business/site owners only offer responsible choices, and we divest from offering convenience junk with high footprints?
- And what more can we be doing with our UX skills to help people understand the problem, or to shift belief? "

ANDREW BOARDMAN, MANOVERBOARD

"The future of the Internet is one color—green. We are all headed down this path, and by hook or by crook, we will realize a culture built on renewables and efficient energy usage. There is no end game. However, what we are building is very large. We need to avoid at all costs the seizure of our liberties by private enterprise and governments. And we need to be very, very wary of virtual reality, despite its cool factor."

CHRIS ADAMS, PRODUCT SCIENCE

"Many of the products we buy we get through some kind of closed loop, product service system, where the incentives for holding onto goods, or returning them for re-manufacture, outweigh the savings from making them disposable. Probably through analytics on their use, or allowing a deeper, two-way conversation with their users about how the current service meets their needs.

"You can look up the likely energy mix of any website and ask the owners why they're still running on fossil fuels.

"We have agreed standards for measuring the impact per visit, or useful work done per kilo of CO_2, and these numbers are exposed through operations dashboards in data centers, and in performance management."

DAVID ANDERSON, CANVAS HOST

"With Greenpeace's estimates that the Internet would be the sixth largest country if based on how much electricity "it" consumes, the metrics show why it's so important to reduce impact. Transparency is the key. With transparency, companies can show what they are doing, and be held accountable for what they are not doing.

Optical computing will be a huge game changer. There will be a tremendous drop in power used. At present, for every Amp used, another is used for cooling. Not so with optical. It will be a tremendous shift from electricity, to space as a precious resource.

Also, to be "green-powered" through any sort of means is now standard. Anyone *not* green powered is on shaky ground.

I think with all current and future datacenters moving to sustainable design and at least renewable power, this could see a tremendous impact right there."

Final Word

From initial concept to completion, this book took about three-and-a-half years to bring to life. Educating myself on the environmental impact of the Internet has changed the way I approach work and my outlook on the role business can have in shaping our future. I hope you found the results useful.

This is a book about the serious threat climate change poses to our planet and the people on it. It is also a book about how those of us on the frontline of building a digital future can make a real and impactful difference.

Thank you for reading it.

[*Appendix*]

Figure Attributions and Links

Throughout this book, I have discussed a variety of companies, organizations, products, and services to illustrate important concepts in and approaches to designing for sustainability. The following list contains relevant links to further information.

FIGURE NUMBER	COMPANY/ATTRIBUTION	URL
p-2	NASA	http://climate.nasa.gov/scientific-consensus
p-4	James Christie	https://docs.google.com/spreadsheets/d/1EV2zavkX487b-3kuHO6C3eQW2WGjjRFSK_WdxJkMfCs/edit#gid=0
p-5	Greenpeace	http://www.greenpeace.org/usa/wp-content/uploads/legacy/Global/usa/planet3/PDFs/clickingclean.pdf
p-6	Tweet Farts	http://www.tweetfarts.com
p-7	Digitalist Magazine	http://www.digitalistmag.com/resource-optimization/2015/12/17/tech-cut-emissions-save-natural-resources-03860595
p-9	US Energy Information Administration	http://www.eia.gov/energy_in_brief/article/renewable_electricity.cfm
1-1	HTTP Archive	http://httparchive.org/interesting.php?a=All&l=Apr%201%202016
1-3	Kelvy Bird	http://www.kelvybird.com/systems-thinking-in-action-2010
1-4	Global Reporting Initiative	https://www.globalreporting.org/resourcelibrary/Informing-decisions,-driving-change-The-role-of-data-in-a-sustainable-future.pdf

FIGURE NUMBER	COMPANY/ATTRIBUTION	URL
1-5	The New Zealand Institute for Crop and Food Research	https://en.wikipedia.org/wiki/File:Helix_of_sustainability.png
1-6	By Zhiying.lim (Own work) [CC BY-SA 3.0 (http://creativecommons.org/licenses/by-sa/3.0)], via Wikimedia Commons	https://commons.wikimedia.org/wiki/File:The_Change_in_Sustainability_Framework.jpg
1-7	AMANAC	http://amanac.eu/amanac-lca-workshop
1-8	Careers New Zealand Government	http://www.careers.govt.nz/practitioners/planning/career-education-benchmarks/revised-career-education-benchmarks-secondary/revised-frequently-asked-questions-about-career-education-benchmarks-secondary
1-9	OnePlanet Sustainability	https://oneplanet-sustainability.org/2013/11/21/corporate-sustainability-profit-motive-and-intention-in-greenwash
1-10	Ecovative Design	http://www.ecovativedesign.com
1-11	Nascent	http://www.nascentobjects.com/#a-shop
1-12	Ecograder	http://www.ecograder.com
1-13	Nest	https://nest.com
1-14	Nest	https://nest.com
1-15	Greenpeace	http://www.greenpeace.org/usa/global-warming/click-clean
1-16	By Sebastian Stabinger (Own work) [CC BY 3.0 (http://creativecommons.org/licenses/by/3.0)], via Wikimedia Commons	https://commons.wikimedia.org/wiki/File:Oculus_Rift_-_Developer_Version_-_Front.jpg
1-17	Google	https://www.google.com/permissions/geoguidelines.html#streetview
2-1	Greenpeace	http://www.greenpeace.org/international/en/campaigns/climate-change/negotiations/COP21-Paris
2-2	Pingdom Tools	https://tools.pingdom.com
3-2	B Corporation	http://www.bcorporation.net

FIGURE NUMBER	COMPANY/ATTRIBUTION	URL
3-5	Green Boilerplate	http://greenboilerplate.com
3-7	Squarespace	https://www.squarespace.com
4-4	Mightybytes	http://www.mightybytes.com
4-5	Google Analytics	https://analytics.google.com
4-6	UX Matters	http://www.uxmatters.com/mt/archives/2012/05/7-basic-best-practices-for-buttons.php
4-7	Optimal Workshop	https://www.optimalworkshop.com/treejack
4-8	StoryStudio	https://www.storystudiochicago.com
4-9	Freitag	http://www.freitag.ch
4-10	Orbit Media Studios	https://www.orbitmedia.com/andy-crestodina
4-11 and 4-12	Hemmings House	http://hemmingshouse.com
4-13	Greenpeace	http://www.greenpeace.org/usa/global-warming/click-clean
4-15	Mightybytes	http://www.mightybytes.com
4-17	Ecograder	http://www.ecograder.com
4-18	Moz Pro Keyword Explorer	https://moz.com/products/pro/keyword-explorer
4-19	Zdeněk Lanc	http://www.slideshare.net/ZdenekLanc/ia-basics
4-20	Tweet Farts	http://www.tweetfarts.com
5-1	Google	https://www.google.com/permissions/geoguidelines.html#streetview
5-3	Mightybytes	http://www.mightybytes.com
5-4	Mightybytes	http://www.mightybytes.com
5-5	Mightybytes	http://www.mightybytes.com
5-6	Mightybytes	http://www.mightybytes.com
5-7	Mightybytes	http://www.mightybytes.com
5-8	Mightybytes	http://www.mightybytes.com
5-9	Apple	iOS screen grab
5-10	Mightybytes	http://www.mightybytes.com

FIGURE NUMBER	COMPANY/ATTRIBUTION	URL
5-11	Mightybytes	http://www.mightybytes.com
5-13	Pingdom Tools	https://tools.pingdom.com
5-15	Adobe	Adobe Photoshop screen grab
5-16	Mightybytes	http://www.mightybytes.com
5-17	Base64	https://www.base64-image.de
5-18	Printfriendly	https://www.printfriendly.com
5-19	Optimizely	https://www.optimizely.com
6-1	drugstore.com	http://www.drugstore.com
	lingscars.com	http://www.lingscars.com
6-2	Pingdom Tools	https://tools.pingdom.com
6-7	WordPress	https://wordpress.com
	Drupal	https://www.drupal.org
6-8	WordPress	https://wordpress.com
6-10	Autoprefixer	https://autoprefixer.github.io
6-11	Mashshare	https://www.mashshare.net
6-12	Disqus	https://disqus.com
	Livefyre	http://web.livefyre.com
6-13	caniuse	http://caniuse.com
6-14	Green Certified Site	http://www.co2stats.com
6-15	Ecograder	http://www.ecograder.com
6-16	Webdev-il	https://webdev-il.blogspot.com/2011/04/what-is-web-accessibility-how-to-make.html
7-6	Hubspot Website Grader	https://website.grader.com
	Ecograder	http://www.ecograder.com
7-13	Greenalytics	http://greenalytics.org
	Green Certified Site	http://www.co2stats.com
	Web Energy Archive Ranker	http://webenergyarchive.com
7-14	Fast Company	http://www.fastcompany.com
7-15	Mightybytes	http://www.mightybytes.com

FIGURE NUMBER	COMPANY/ATTRIBUTION	URL
7-17	Climate Ride	*http://www.climateride.org*
7-18	Ecograder	*http://www.ecograder.com*
8-2	B Impact Assessment	*http://bimpactassessment.net*
8-3	Green Web Foundation	*http://www.thegreenwebfoundation.org*
8-5	Fairphone	*https://www.fairphone.com*
8-6	Udemy	*https://www.udemy.com*

[Acknowledgments]

I interviewed more than two dozen people during the process of writing this book. These leaders were very generous with their time and provided insights that helped me tell a much better story.

B Corp Leaders Profiled in This Book

More than a dozen of the people interviewed either own or work for certified B Corporations, or B Corps, which is why you read a lot about B Corps in this book. They are leading a revolution to use business as a force for good in the world. Supporting these businesses means supporting a future with a shared and durable prosperity for all. In addition to their bios, they also shared why being a B Corp is important to them.

ANDREW BOARDMAN

Founder and principal, Manoverboard

Manoverboard is a B Corp that helps socially responsible businesses, educational institutions, and large nonprofits to build and motivate their audiences. The company uses design as the primary tool in a strategic communications tool belt to help organizations promote their message, create conversations, and connect with audiences. Manoverboard has designed digital solutions and identities for Generation Investment

Management (former Vice President Al Gore's investment firm), RAND Corporation, MIT, Acumen Fund, United Nations University, Council of Canadians with Disabilities, Maestral International, University of Manitoba, SJF Ventures, and Encourage Capital. The company starts by asking hard questions to get to the very heart of an organization's goals and commitments. We then work to boil down ideas and messages to build compelling identities, websites, and marketing tools. The key is to create lasting and durable design solutions for our clients that help them reach farther and connect globally.

Why being a B Corp matters

> Running a business and trying to impact the world in a positive way can be at times like mixing oil and water. B Corp helps me fulfill two important goals as a business owner. First, it concretizes my personal values and beliefs within a reliable, measurable, and rational structure through B Lab's Impact Assessment. Second, as a B Corp, Manoverboard makes a public statement along with thousands of other businesses that capitalism must evolve and take on the hard work confronting us—including solving many of the social and environmental issues that capitalism has helped create. B Corp marks a conscious change in the language, culture, and practices of business.

MIKE GIFFORD

Owner, Open Concept Consulting

Mike Gifford is the founder and president of OpenConcept Consulting Inc., a web development agency specializing in the Drupal software, and a Benefit corporation. The company has more than 16 years of experience in designing secure, scalable, inclusive, and sustainable solutions for the non-for-profit, public, and private sectors as well as an historic engagement in striving to build a better Internet using open source software.

Pushed by a desire to build better and more inclusive software, Mike has been involved with accessibility since the beginning of the 90s. He has spearheaded accessibility improvements in Drupal since 2008, and officially became Drupal's Core Accessibility Maintainer in 2012. As a techie at heart, Mike likes to get into the code when he gets the chance. Being familiar with everything from accessibility issues to system administration provides the ability to understand the technological big picture.

Why being a B Corp matters

> I've found being a B Corp increasingly important. The impact assessment process is challenging and makes me reflect on how my business can do better. There really are so many things that we can do to make our operations better, and so rarely is there time for that reflection. There are thousands of ways to reduce our impact on the planet, touching base with others who have similar goals helps inspire me to keep looking at what we should do next. By working together there are opportunities to save time and money, too.
>
> More importantly, I am inspired by the possibilities of working together with other B Corps in order to change the much bigger problem of climate change. As Naomi Klein highlighted in *This Changes Everything*, our society is going to have to change radically if we are to rapidly shift to reduce our CO_2e emissions. It can seem hopeless working in isolation, but being part of a network spanning the globe and full gamut of businesses it seems more possible. Capitalism needs to shift to be not only for profit, but to serve the broader goals of society. We all need a healthy environment, a vibrant community, and a positive workplace. Business needs to play a role in making this happen. B Corps are one type of social enterprise moving us in that direction.

ROBERT STEVENS

Head of partnerships, ClimateCare

ClimateCare is a certified B Corp with a target to improve 20 million lives and cut 20 million tons of CO_2e by 2020. Robert Stevens, joined ClimateCare in 2007 and oversees its partnership programs—engaging with corporates, governments, and entrepreneurs to meet their business, environmental, and social development goals. Rob ensures that ClimateCare programs work for all partners, are outcome-focused, and deliver measurable results for business, people, and the environment. Rob lives in Oxford, England, with his wife and two children. He holds a degree in business administration and is currently studying a postgraduate certificate in sustainable development with the School of Oriental and African Studies at the University of London. Rob is also chair of the International Carbon Reduction and Offset Alliance (ICROA).

Why being a B Corp matters

> We have been a triple-bottom-line company from the beginning, self-described as profit for purpose. Becoming part of a movement of like-minded companies has given us a clear, succinct way of describing what we want to achieve. In August of 2015, we became a founding UK B Corp with a score of 141 (which is high in B Corp terms). The assessment gave us clear ways to identify strengths and areas for improvement and it helped us to get our team motivated and engaged around a common set of goals. We have also found the B Corp community to be a great network of companies which has allowed many collaborations and partnerships to emerge.

JEN BOYNTON

Editor-in-chief, Triple Pundit

Jen is editor-in-chief of TriplePundit, an online community dedicated to furthering the conversation on triple-bottom-line business. She has an MBA in Sustainable Management from the Presidio Graduate School and volunteers with Defy Ventures, an entrepreneurship, employment, and character training program for people with criminal histories. At TriplePundit, she's had the pleasure of working with brands including PwC, SAP, CVS Health, Kimberly-Clark, H&M, Adobe, Levi Strauss & Co., and Yum! She lives in San Diego with her husband and toddler overlord. Hit her up on twitter @jenboynton to discuss picky eaters, the prison industrial complex, or sustainability reporting methodology.

Why being a B Corp matters

> It always gives us new ideas about how to improve and become a better company while making an impact, too.

MIQUEL BALLESTER SALVÀ

FAIRPHONE

Product manager and innovation lead, Fairphone

Miquel Ballester Salvà is cofounder and responsible for product management and innovation at Fairphone, a social enterprise which uncovers complex production systems to change how things are made. Fairphone opens up the supply chain of a mobile phone, developing projects with suppliers, consumers, and other influencers to create a fairer economy and improve the life cycle. After graduating in Industrial Design Engineering specializing in technologies for sustainable development, Miquel has contributed to Fairphone in both product and system design. He participated in the exciting process of turning this organization from a nonprofit to a social venture.

Why being a B Corp matters

> We knew about B Corps early on. In general, we have been very critical of "green labels," but becoming a B Corp helped us formalize more socially and environmentally conscious practices. It also gives us a great opportunity to look in the mirror every year and reevaluate the kind of company we want to be. Plus, the B Corp network is wonderful.

SHAWN MILLS

President and CEO, Green House Data

Shawn Mills is president and chief executive officer at Green House Data as well as a founding member of the company. Under his leadership, Green House Data has undergone rapid expansion since 2007, launching cloud and colocation data centers, coast to coast. Shawn has been chosen to speak at industry events and locations like Data Center World, the National Center for Super Computing Applications and Cloud Computing Expo, and his writing has appeared in EdTech Magazine, Data Center Knowledge, and Manufacturing Digital, among other publications.

Why being a B Corp matters

> It's a very supportive community that shows us better ways to be sustainable throughout our offices and daily work routines. It serves as a benchmark and a way to show the outside world that we have met a certain standard and aren't just claiming to be making these efforts. Since we certified as a B Corp we have redoubled our commitment to efficiency and renewables as a core part of the business. This was always our business model, but adding that B Corp stamp means we have woven it further into our identity. It's kind of an ever-present reminder of what we are working towards. With recertification coming up we need to continue to find ways to improve, too, especially as the company continues growing.
>
> The B Corp movement has also encouraged us to practice corporate sustainability beyond the environmental angle. A commitment to our financial sustainability, our community involvement, and our employee well-being means we have happier workers, investors, and customers. I run Green House Data in as transparent a manner as possible, walking

the entire company through our financials and the road map I have in mind for our future at regular meetings. Our management takes an open-door policy, meaning even the interns can pop their head in my office any time. Being a B Corp has solidified these practices and held us accountable to them.

EMILY LONIGRO BOYLAN

President, LimeRed

In 2004, Emily founded user experience (UX) agency LimeRed Studio with a plan to prioritize high-quality design, user experience, and meaningful social impact. Lime Red is now the nation's only WBE certified B Corp UX agency. Over the past 12 years, she's built this business into a presence with national clients focused on doing work that has meaning.

Her experience spans design, business development, UX, writing, marketing, and strategy in both online and offline programs for multinational corporations, nonprofits, universities, boutique businesses, and prestigious consumer brands. Today, LimeRed's focus is on creating stunning creative pieces for nonprofits, conscious companies, and educational organizations to do work that makes people's lives better.

She is vice president of the board of nonprofit MOM+BABY, member of the Social Enterprise Alliance, member of Ms.Tech's Executive Circle, and active in NTEN. She's run accelerators for nonprofits and is a mentor for other women in business. She's a published writer and teaches workshops on business development, information architecture, UX design, branding, and content strategy. Emily is also a regular speaker and panelist at national and regional conferences.

She's a mom, activist, small business owner, designer, and activator.

Why being a B Corp matters

> Being a B Corp has fundamentally changed my business in more ways than I can explain. Every decision we make in L.A.B. is through a B Corporation lens—for us and for clients. We became a B Corp at the end of a particularly hard year. Going through the assessment gave me a different perspective and hope for what my business could be. We had always been walking down the do-good road, and having that certified felt great. The community is inspiring: I've met some amazing business owners who I trust completely. It helped me personally focus on the big picture when the daily grind seems overwhelming. It helped me embrace delegation. It helped me reconsider how I hire and how I reward my staff. It reminds me that we all have different motivations and reasons for showing up at work. At the time we certified, after 10 years and a lot of tough decisions, it made me fall back in love with my business again.

GREG HEMMINGS

 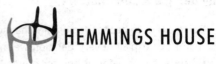

CEO, executive producer, Hemmings House

Filmmaker Greg Hemmings leads from the heart, making the world a better place one story at a time. His work connects individuals and the world, celebrating the universal in the particular, and exploring how our collective human narrative is unfolding against the unique backdrop of these interesting, accelerated times.

Greg's quest has taken him around the world to tell global stories that inspire local change and share local tales with international resonance. More than 60 broadcasters have aired his work, which includes TV series and documentary films that focus on social justice, collaborative problem solving, and sustainable growth. His work is sensitive, honest, and inspiring—kind of like Greg himself.

You see, Greg doesn't just make films about other people making positive social change; he lives this ethos, too. His companies, Hemmings House Pictures and FYA.tv (For Your Action), are certified B Corps, a progressive, triple-bottom-line model that sees business as a force for good by valuing people and the planet as much as profits. Greg leads a team of change-makers who share his vision of making films that make a difference as well as his commitment to volunteerism and giving back.

Greg is himself building a movement of change, empowering others to become leaders in their own ways, in their own communities. He has mentored numerous emerging entrepreneurs and artists. His weekly podcast, The Boiling Point, celebrates positive change in business and the world. And he is an in-demand speaker about the links between documentary filmmaking, entrepreneurship, triple-bottom-line business, and social movements. Greg is increasingly asked to share his views on how companies can inspire positive change and increase brand trust trough social-impact films.

Why being a B Corp matters

> Being a B Corp has introduced us to a whole new world of understanding. The concept of green hosting never occurred to us before we met Tim Frick at the B Corp Champions Retreat. Having the positive peer pressure in the B Corp network really helps us and encourages us to invest in positive changes in our company. Hemmings House has always been a sustainable company, but being a B Corp is now helping us to go further. The measurements and the personal competition against our own scores helps us think creatively as well about ways we can do better next assessment round. We are now held at a much higher standard.

ZACH BERKE AND PHILLIP CLARK

Chief executive officer, chief creative officer, Exygy

Zach is the Founder and CEO of Exygy, a software strategy, design and build agency based in San Francisco. Exygy works with the world's leading change-makers and innovators, using technology as an amplifier for change. As CEO, Zach is responsible for steering the 15-person company, leading strategy work with clients to shape vision, and ensuring world class execution by Exygy staff. Zach has led a certified B Corp since 2010 and is a local leader in the San Francisco Bay Area B Corporation community. Zach has directed Exygy's work for clients including Google.org, the United Nations, Yahoo!, The City of New York, The City and County of San Francisco, UNICEF, The UN World Food Program, Skoll Foundation, Zendesk, Intuit, EMC, and many others. Zach is the proud father of three little dudes and proud owner of three big bicycles. Zach and his wife Dr. Gabriella Bartos reside in the Outer Lands of San Francisco.

Phil graduated Cum Laude from Vassar College with an AB in media studies and earned a master's in science of teaching from Pace University. As chief creative officer for Exygy, Phil works with Zach to set the overall vision, processes, and strategic direction of the company. A triathlon enthusiast, Philip lives in Oakland where he can swim, bike, and run in a sunnier clime. He sometimes wishes he were in Cinque Terre in Italy, his favorite place to visit.

Why being a B Corp matters

> Being a B Corp supports our mission to design exceptional software for the world's leading change makers. It aligns our team, clients, and partners with a common purpose of improving our community and environment.

DAVID ANDERSON

Owner, Canvas Host

David is the owner of Canvas Host, a Portland, Oregon domain registrar, WordPress development, and cloud hosting service provider. From an early age, he was taught and mentored to leave a system in better shape than he found it. His company is relentlessly committed to environmental responsibility and corporate leadership, echoed in its B Corporation certification, ongoing sustainability program, and participation in renewable energy programs, community stewardship, and nonprofit fundraisers. A cornerstone of his personal and company objectives is to use technology in a way to benefit others, to help green the environment, and serve as an example of what responsible business can look like. When applied with the right focus and direction, business truly can be used as a force for good.

Why being a B Corp matters

> It compels us to always look forward to a better Canvas Host, to never stop challenging us to improve our business, or make our impact a positive one on our community and clients. I could easily go on for pages about the ways being a B Corp has helped us, so instead I will summarize that our sustainability program sprang from a desire to achieve transparency in all things, and as a B Corp, we have taken it to increasing levels and seen grand achievements in our processes, our metrics, and the way our own customers engage with us.

EMILY UTZ

Director of operations, DOJO4

Emily Utz is DOJO4's sustainability consultant. She now applies more than a decade of work for human rights and environmental nonprofits to helping businesses become catalysts for positive change, socially and ecologically.

Why being a B Corp matters

DOJO4 is not really a tech *shop*. We're a group of people who work on projects in the realm of technology, art, social change and business. While we are very proud of the code and design that we put out in the world, what we value most is our humanness. We run a healthy business in order to support the people who work here, our families, and our community. Becoming a B Corp was a natural evolution of our commitments to people and place, extending that globally to ensure that the clients we work with and the products and services we procure are aligned with our efforts to make a positive impact in the lives of everyone we interact with.

ANDY CRESTODINA

Partner and strategic director, Orbit Media Studios

Andy is a cofounder and the strategic director of Orbit Media, an award-winning 38-person web design company in Chicago. Over the past 15 years, he has provided web strategy and advice to more than a thousand businesses. As a top-rated speaker at national conferences and as a writer for many of the biggest blogs, Andy has dedicated himself to the teaching of marketing.

Andy has written hundreds of articles on content strategy, search engine optimization, social media, and Analytics.

- *Forbes* Top 10 Online Marketing Experts to Watch in 2015
- *Entrepreneur Magazine* Top 50 Marketing Influencer in 2016
- Mentor at 1871, the #1 incubator in the United States
- Adjunct professor of digital marketing at Loyola University

He is also the author of *Content Chemistry: The Illustrated Handbook for Content Marketing.*

Why being a B Corp matters

> It means raising our hand and saying that business is part of a bigger picture. It means that business, society, and the environment all fit together, work together. It's a way to make the phrase 'work-life balance' official in our company. We are capitalists, but we believe in sustainability in all things, for the environment, and also for the economy and for lives of the people on our team.

JD CAPUANO

Cofounder, Closed Loop Advisors

JD solves business problems. He works and teaches at the intersection of strategy, data, technology, and responsible business. His history of project implementations informs his strategic advice to walk the line between what's practical and innovative. JD has a diverse history of roles ranging from being an entrepreneur at a medium-sized company to being in a management position at a Fortune 200 company to being an entrepreneur and consultant. He has worked for or advised organizations in healthcare, business services, ecommerce, high tech, furniture manufacturing, non-profit, and government. He co-founded Closed Loop Advisors, a Certified B Corp. JD also teaches at Bard College's MBA in Sustainability. He earned an MS in Sustainability Management from Columbia University and a dual BA in Business Administration and Interdisciplinary Studies from the University of Pittsburgh.

Why being a B Corp matters

B Corps demonstrate that companies can be successful while also being responsible with their staff, community, and the planet. What I really like about the B Corp movement isn't just that it guides and motivates individual companies to be more responsible, rather that it symbolizes an evolution in capitalism to be more socially and environmentally responsible. I also really like how B Lab collects a lot of data that proves financial performance and responsibility can go hand in hand.

JILL POLLACK

Founder, StoryStudio Chicago

As chief story wrangler, Jill Pollack spends her time chasing down the best stories...and making them better. But telling a great story isn't enough for her; she has to throw some neuroscience into the mix. Art + Science = the answer to everything. Jill is the founder and director of StoryStudio Chicago, a writing training center for creative writers and business professionals. In addition to teaching, writing, and forcing people to admit that they can't live without great stories, Jill oversees writing training for more than 1,200 students each year.

She is a frequent speaker on the power of stories in our personal and professional lives and was once again included in the Newcity Lit Top 50 list of literary leaders in Chicago.

Why being a B Corp matters

It's all about community. What communities we become a part of, what we give to those communities, and what we receive in return.

StoryStudio's stakeholders may not come from the same place geographically, but we are all part of a larger, worldwide writer's network. Our mission includes connecting our students with this network and the opportunities it provides. For me personally, it's the warmth, knowledge, and support I receive from our local B Corp community. I remember receiving our certification and then immediately being adopted by this vibrant group of business owners and entrepreneurs. I count them as friends and advisors and can't imagine running my business without them.

NOEL BURKMAN

Partner, Month16

After earning a degree in computer science from DePaul University, Noel decided to head west, landing in the Bay Area. It was in the dot-com epicenter that he launched the first of many startup ventures he would lead over the years.

In 2000, he left the entrepreneurial world for some corporate tempering at Follett Higher Education Group. While at Follett, he developed an ecommerce team and within four years led the company from zero ecommerce revenue to more than $300 million a year in ecommerce revenue, establishing Follett as one of the top 75 revenue generating e-commerce sites in the world. He then spent the next several years incorporating retail, wholesale, and digital media into their overall strategic plan.

Noel's next adventure took him to Washington, DC, to work at the American Press Institute (API). As the director of digital transformation, he advised and guided senior leadership from various media companies including the *Tribune*, *Washington Post*, and Gannett on transforming the role of legacy media companies (in other words—print newspapers) by developing and implementing new business models.

There's a line in *The Godfather* that goes "Just when I thought I was out, they pulled me back in." In 2010, Noel returned to Follett, spending the next three years developing and implementing a new brand and business model for the 140-year-old company. In the summer of 2011, he developed and led the Skyo.com campaign, which was launched by Follett to combat its rapidly eroding online market share. Although the following is unverified, they were pretty sure Skyo.com was the fastest growing ecommerce site, going from the first line of code to more than $10 million in revenue in less than 9 months.

In early 2014, he joined the executive leadership team of a Chicago-based luxury chocolate company and faced myriad challenges including brand confusion, competitive onslaught, and an inefficient patchwork technology infrastructure. He worked with an amazing team of people to develop both brand strategies and a technology architecture to support a high-touch brand in an online world.

Currently, Noel is a cofounder at Month16, which is also his first Benefit Corporation. Month16's mission is to increase the ROI of early stage and declining business in a conscious way.

Why being a B Corp matters

> Have you ever heard the phrase "It's just business"? If so, most likely someone just got screwed out of something, and the screwer is rationalizing an action that is morally void. To me, the B Corp is about culture, balance, and long-term value. I am a capitalist through and through. That said, it matters to me how profits are derived and with what impact. The fact of the matter is that a business with a B culture will drive more value over a longer period of time versus a business that is solely focused on profitability—with happier and healthier employees, supply chains, partners, and customers.

OTHER B CORPS

Also, thanks very much to Ryan Honeyman, author of *The B Corp Handbook*, and the editorial team at B Corp Berrett-Koehler Publishers for providing me initial feedback and guidance on early drafts of the outline.

NGOs, Educators, and Design Leaders Profiled in This Book

DAVID POMERANTZ

Senior climate and energy campaigner, Greenpeace USA

David Pomerantz is the executive director of the Energy and Policy Institute.

Prior to joining EPI, David spent eight years working with Greenpeace to move the electric sector away from fossil fuel and towards renewable energy. As a senior climate and energy campaigner, David coordinated Greenpeace's *Clicking Clean* campaign for a greener internet. The campaign resulted in major internet technology companies, including Apple and Amazon, committing to power their data centers with 100% renewable energy. The campaign catalyzed hundreds of megawatts of new renewable energy development, and helped to defend and expand renewable energy policies and investment in the United States and abroad. David also led the development of Greenpeace's strategy to push utilities toward renewable energy in Iowa, Nevada, North Carolina, and Virginia.

In addition, David designed and implemented the communications strategy for Greenpeace's work on electricity, leading to media coverage in Bloomberg Businessweek, *Los Angeles Times*, *The New York Times*, *USA Today*, *Wall Street Journal*, *Washington Post* and *Wired* magazine. He also helped to build Greenpeace's field presence on coal-related issues in 2010 and 2011 and worked with communities in Ohio and New England in 2008 and 2009 to push members of Congress to support climate change and toxic chemical legislation. He previously worked as a reporter for newspapers in Boston and New York City.

David graduated Summa Cum Laude with a degree in History from Tufts University. A transplanted New Yorker, he now lives in San Francisco.

PETE MARKIEWICZ

Author and instructor, The Art Institutes

Pete's background is in science. He originally graduated with a doctorate in molecular and cellular biology from the University of Chicago. Later, he researched multivalent vaccines, and ultimately landed at UCLA, where he studied protein structure and evolution under Dr. Jeffry Miller. In late 1993, Pete made a huge career shift from biology to interactive design and development. He left UCLA to cofound Indiespace (formerly Kaleidospace) with Jeannie Novak. Indiespace went online in March 1994, and was the first web-based arts and entertainment company to sell indie products. During the 1990s and early 2000s, Indiespace was the prototype for modern video and music streaming websites. The site is still available today at *http://www.indiespace.com*. In addition to Indiespace, Pete also cowrote three books on the Internet, technology, and entertainment with Jeannie Novak, which are available on Amazon.

Pete has a strong interest in generations, and beginning in the early 2000s, worked with William Strauss and Neil Howe of Lifecourse Associates, the creators of the "Millennial" generation concept. He coauthored its book *Millennials and the Pop Culture* in 2006, which predicted many of the trends we see in millennials today. He also developed seminars on the Millennial Generation, pop culture, and virtual worlds for USC's CTM Programs at the Marshall School of Business.

In 2005, Pete was a team leader for Team Robomonster, a robotic, self-driving rock vehicle entered in the DARPA 2005 Grand Challenge. The team, composed mostly of web design and development students, made it to the second of three rounds in the contest. Currently, Pete teaches interactive and web design at the Art Institute of California, Los Angeles, and consults with web, game, and virtual reality companies.

In the last few years, Pete's web experience led him to become concerned with the long-term sustainability of the online world. His main interest is extending web sustainability beyond web performance (WPO) alone, to create a framework for sustainability modeled on those found in other fields including architecture and industrial design. Pete has also been developing "Green Boilerplate"—a concept for a sustainable web architecture that can be used as a starting point for web-based projects.

JAMES CHRISTIE

mad*pow

Director, experience design, Mad*Pow (Portsmouth NH)
Cofounder of Sustainable UX: Digital Design versus Climate Change Conference

Based in New Hampshire, James is a user experience design director at MadPow.com, an experience design agency known for its work in the financial, education, and healthcare spaces. James has been writing and presenting about sustainable web design since 2012. His article

"Sustainable Web Design" appeared in issue 383 of A List Apart in 2013. He is the organizer of Sustainable UX, a virtual conference for digital designers who care about climate change.

ERIC JANOFSKI

Founder, Base1

Hailing from the Upper Peninsula of Michigan, Eric Janofski is a seasoned web developer and small business owner. He loves warm summers, short winters and spending time outside with his son.

RENÉ POST

Cofounder, The Green Web Foundation

René Post has pioneered Internet organizations since the mid-90s centered around the keywords transparency, empowerment, decentralized, sensor networks with a special focus on low-tech solutions for modern problems. He is the cofounder of The Green Web Foundation (*www.thegreenwebfoundation.org*).

JOHN HAUGEN

Cofounder and sustainability advisor, Third Partners, LLC

John produces solutions to key sustainability challenges for organizations of all types. He develops programs such as carbon footprint management, green building strategies, and portfolio-wide energy analytics. He holds a BA from Illinois Wesleyan University and an MS from Columbia University. John also mentors startup companies through Cleantech Open, the world's largest cleantech startup incubator.

CHRIS ADAMS

Director, Product Science

As an environmentally focused tech generalist, Chris Adams worked the past 15 years as a designer, user researcher, sysadmin, product manager, and developer with startups, NGOs, blue-chip companies, and UK government departments before setting up Product Science Ltd, a digital agency in 2013.

TODD LARSEN

Executive co-director, Green America

Todd Larsen is the executive co-director for Green America. He directs Green America's corporate responsibility and consumer engagement programs. Green America's corporate responsibility programs educate consumers and investors about the environmental and social records of major companies and encourage them to take action to promote greater responsibility. In addition, Green America provides tools and resources to companies to help consumers, businesses, and investors improve their impacts on people and the planet. Todd also heads up Green America's Climate Solutions Program which works to end polluting energy sources, map the future for clean energy, and engage Americans as investors in a clean energy future. Todd is also a Senior Fellow, Green Consumer Trends, with Green America's Center for Sustainability Solutions, where he provides the latest market research and consumer purchasing data to companies to help them see the demand for more responsibly sourced products and services. Todd has 15 years of experience in public education and corporate campaign work and has a master's degree in Political Science from the University of Wisconsin—Madison.

Many thanks to Ian Jacobs of the World Wide Web Consortium, Greg O'Toole, instructor at Penn State University and author of *Sustainable Web Ecosystem Design*, Kem-Laurin Kramer, author of *User Experience in the Age of Sustainability*, and Dave Bevans for the incredibly helpful feedback on my initial outline. Thanks as well to the talented editorial team I worked with at O'Reilly—Angela Rufino and Nick Lombardi—for helping me transform a rough outline into a cohesive narrative.

Team Mightybytes

Writing a book takes up a huge amount of time. The process is exciting, unnerving, all-consuming, rewarding, and frustrating. It requires patience, tenacity, and the willingness to give up things in life and in work to which you have become accustomed.

Although this was my fourth time down the book writing path, it was probably the most difficult for several reasons. One, it was a true passion project, so I really threw my all into it. Two, the timeline was tighter than other books and required more research and more attention over a shorter period of time. Three, it came at a time when Mightybytes was executing several very large, ongoing projects, which proved challenging for the business, the book, and my relationships with coworkers.

I must express heartfelt thanks to the team at Mightybytes who either helped with this project or held down the fort during the writing process. Thanks to Amber Vasquez and Eric Mikkelson, who let me interview them for the book, plus extra thanks to Carl Baar for all the help with figures, images, and illustrations.

And finally, to my supportive family and my partner in life and in crime, Jeff Yurkanin, I am so very grateful. Your support is so appreciated and I love the madcap life we have built together.

[Index]

A

A/B testing, 113, 138, 180
Accenture, xxii
accessibility
   performance optimization and, 225–227
   standards for, 42, 97, 183–184
   UX considerations, 181–182
   video and audio, 129
Adams, Chris
   about, 313
   on estimating carbon footprints, 240–241
   on green hosting, 269
   on safe disposal, 29–30
   on sustainable Internet, 283–284
   on workflows, 88
Adobe Illustrator, 160
Adobe Photoshop, 175
Adobe Typekit, 172–173
Advanced CSS/JS Aggregation module (Drupal), 208
Agency Spotter, xxii
Agile methods
   about, 90–91, 129–130
   Ecograder case study, 250–251, 254–255, 257
   lesson from, 135–139
   performance optimization and, 221–222
   preparing over planning, 133–135
   sprints, 91, 221–222, 250–251, 254
   user experience and, 157
   user stories, 164
The Agility Shift (Meyer), 130–131
agriculture and food production industry, xxii–xxiv
AIGA, 272
Airbnb, 96
airline industry and GHGs, 3
AISO.NET, 78

A List Apart, 186, 192
All in One WordPress Security and Firewall plug-in, 204
Amazon
   open source practices, 96
   performance calculations, 191
   renewable energy and, 72, 147
   video streaming, 34
   VR content, 37
Amazon Web Services. *See* AWS (Amazon Web Services)
American Press Institute (API), 307
AMP project, 147, 205
Anastas, P.T., 166–167
Anderson, David, 71, 267, 284, 302–303
Andreessen, Marc, xxi
Apache Soir module (Drupal), 209
API (American Press Institute), 307
Apple Computers
   Flash support and, 250
   green hosting and, 266
   Homekit framework, 31
   powering data centers, 40
   renewable energy and, 147
   sustainability challenges, 67–68
   UX considerations and, 188
appliance and equipment standards, xxv
Art Institutes (California), 93, 310
Atticus Digital, 278
auditing content, 114–118
Authenticity Checker plug-in, 205
Autoprefixer, 215–216
AWS (Amazon Web Services)
   green hosting challenges, 40, 68–69
   renewable energy and, 65, 105, 164
   social media strategies, 145
   video streaming and, 35, 65

B

bandwidth considerations
  Flash and, 188
  mobile-first strategy, 185
  performance optimization, 192, 197
  progressive enhancement, 42
  responsive web design and, 186–187
  shared resources, 248
  video compression, 128–129, 192
  video streaming, 34–35, 108, 124–126
  VR content, 38–39
Bartos, Gabriella, 302
Base64 encoding, 172, 177–178, 220
Basel (company), 312
baseline for benchmarking, 15
B Corporations
  about, xxix–xxxi, 103, 291
  future-friendly Internet and, 264–266
  stakeholder model and, 85–86
Belgium, e-waste and, 272
benchmarking
  about, 19
  content strategies, 114, 136
  Ecograder case study, 255, 257–261
  setting baseline, 15
Berke, Zach, 157, 301–302
Bevans, Dave, 314
B Impact Assessment, xxix, xxxii, 22
biomimicry process, 22
BJ Lazy Load plug-in, 204
B Lab, 77, 80, 292
Blueprint framework, 92
Boardman, Andrew
  about, 291–292
  on content strategy, 109
  on future-friendly Internet, 283
  on green hosting criteria, 104
  on service reliability, 81
The Boiling Point podcast, 301
Bonneville Environmental Foundation, 72, 267
Boost module (Drupal), 209
Bootstrap framework, 92
Boynton, Jen, 145, 147, 295
branding in business, 20–21
browsers
  carbon footprints, 247
  CSS sprites and, 176–177
  performance optimization, 194–195, 204–205, 208–216, 219–222, 227
  print styles and, 178–179
  progressive enhancement, 185
  responsive design, 186
  web standards and, 97, 184
Brundtland Commission, 4–6
BSR (Business for Social Responsibility), xxxii
Burkman, Noel, 275, 307–309
Burns, Thomas, 100
business sustainability
  about, 9–12
  benchmarking, 19
  branding, 20–21
  greenwashing, 20–21
  identifying efficiencies, 12
  innovation and disruption, 22–25
  life cycle assessments, 12–18, 26–27
  marketing, 20–21

C

C2C (cradle-to-cradle) assessments, 9–11, 12
Cabot Creamery Cooperative, 87–88
Cache-Control header, 210–211
CakePHP framework, 92
calls-to-action (CTAs), 114, 132–133, 138
caniuse.com, 219
Canvas Host, xxix, 72, 81, 302–303
Capuano, JD, 236–237, 21, 305, 25–26
carbon dioxide equivalent (CO2e)
  about, xviii
  environmental impact example, xvi–xvii
  Internet usage and, xx–xxi, xxvi–xxvii
carbon footprints
  Apple Computer, 67–68
  calculating, 229–233
  of common tasks, xvi–xvii
  Ecograder case study. *See* Ecograder
  framework for estimating, 234–240

of online searches, 140
of running websites, 25–26
sustainable web design and, 55
of 3D printers, 24
Carbon Management, 230–231
Cascade framework, 92
CDN module (Drupal), 209
CDNs (content delivery networks), 43, 26
Center for Biological Diversity, 71
certification, 15, 85
Chase, Robin, 96–97
Chicago Sun-Times, 255
Chouinard, Yvon, 7
Christie, James
   about, 281–282, 311
   on carbon cost of research, 158
   on carbon footprints, xvi–xvii, 234–236
   on page bloat, 192, 217
   on social media sharing plugins, 216
   on sustainable Internet, 281–282
Cisco Visual Networking Index, 35
Clark, Philip, 157, 301–302
Cleantech Open, 313
Click Clean Scorecard, xx, xxix
Clicking Clean annual report
   on Apple Computer, 67
   on environmental impact of Internet, xviii–xix, 35–36
   on green hosting challenges, 103–104
   on political challenges, 71
   on social networks, 145
Climate Action Plan, xxv
ClimateCare, 76, 294
climate crisis
   B Corporations and, 293
   climate versus weather, xiii–xiv
   connecting the dots, xv–xvii
   Goliath/Frank Winter Storm, ix–xii
   impact of an individual on, xvi–xvii
   MadPow and, 281–282
   open source projects and, 96
   Paris Agreement on Climate Change, 51
   renewable energy credits and, 74
   Squarespace and, 101
   sustainable mission statements on, 87
   UN Conference on Climate Change, 49
Climate Ride, xvi–xvii, 135, 257–261
Closed Loop Advisors, 305
closed loop systems, 9–10
CMS (Content Management Systems), 97, 201–209
CO2e (carbon dioxide equivalent)
   about, xviii, 230
   environmental impact example, xvi–xvii
   estimating, 234
   Internet usage and, xx–xxi, xxvi–xxvii
CO2Stats.com tool, 223, 241, 252
CodeIgniter framework, 92
code obsolescence, 45–46
collaboration in design teams, 112, 155–156, 196–197
color choices (visual design), 169–170
comments, performance optimization and, 207, 211, 217–218
component design, 163
cone of uncertainty, 89
Conscious Capitalism, 264
construction industry, xxii–xxiv
content audits, 114–118
Content Chemistry (Crestodina), 111, 305–306
content delivery networks (CDNs), 43, 25
Content Management Systems (CMS), 97, 201–209
content patterns, 162–163
content strategy
   about, 110–111
   Agile methods, 129–139
   avoiding dark patterns, 164
   CDNs and, 25, 43
   component design, 163
   content audit, 114–118
   content conundrum, 107–110
   content patterns, 162–163
   defining the rules, 112–120
   display patterns, 163
   Ecograder case study, 254
   information architecture, 118–120
   measuring performance, 113–114

page briefs, 162–163
potential barriers and workarounds, 147–148
storytelling in, 120–123, 164
sustainable searches, 139–146
user experience and, 112, 162–167
video considerations, 124–130
web design considerations, 59–60, 63–65
Content Strategy for the Web (Halvorson), 110
conversion optimization, 113
Cook, Gary, 1
cookies, performance optimization and, 212
COP21, 50–51
corporate culture (business function), 10
cradle-to-cradle (C2C) assessments, 9–11, 12
cradle-to-grave assessments, 11, 12
Creative Bloq, 41
Crestodina, Andy
    about, 304–305
    on costs for online actions, 146
    on search engine optimization, 122, 141
    on sustainable content, 111
    on video content, 124, 129
CRM (Customer Relationship Management), 197
Croll, Alistair, 136
CSA Group, 15
CSS/CSS3 standards
    about, 42, 92
    accessibility and, 183–184
    adoption of, 222
    Ecograder tool and, 245
    embedded fonts and, 172
    image formats and, 174
    performance optimization, 195, 201, 204, 208–216, 218–219
    print styles and, 178–179
    responsive design and, 186–187
CSS sprites, 176–177, 210, 213, 221
CTAs (calls-to-action), 114, 132–133, 138
curation, content, 137
Customer Relationship Management (CRM), 197

customer service (green hosting), 101–104

D

Daniel, Jennifer, 51
Dark Patterns website, 165
Data Center Infrastructure Efficiency (DCIE), 267
data centers
    emissions from, xx, 26–27
    green hosting, 266–268
    hardware considerations, 270–271
    power usage by, xxii, 38–40, 74
    renewable energy and, xxii, 30, 67–69, 75–79, 105
    SEO and, 141
    social strategies, 145–146
    sustainable web design and, 152
data disposal, 45–47
data interpretation (LCAs), 12–13, 18
Davis, Jed, 88
DB Maintenance module (Drupal), 209
DCIE (Data Center Infrastructure Efficiency), 267
Delicious, 109
dematerialization strategy, 151–152
Department of Energy (DOE), xxv, 269
Design is the Problem (Shedroff), 6, 54, 151–152, 230
"Design through the Twelve Principles of Green Engineering" (Anastas and Zimmerman), 166–167
Dieselgate, 20
digital carbon footprints. *See* carbon footprints
Digitalist Magazine, xxii, xxiv
Digital Reality, 36
discovery workshops, 164
display patterns, 163
disposal
    of hardware, 47
    as LCA component, 17, 41
    as virtual LCA, 44–46
Disqus comments system, 217
disruption (business processes), 22–25

distribution (LCA component), 17, 41
distribution, video content, 128
DNS lookups, 211–212
Document Object Model (DOM), 212–213
DOE (Department of Energy), xxv, 269
DOJO4, xxix, 303–304
Dolphin Blue, 87
DOM (Document Object Model), 212–213
Dropbox, 35, 239
Drupal software
   about, 208–209, 292–293
   CMS usage, 201, 203
   search support, 143
Ducker, Michael, 147

E

Ebay, 36
Ecograder
   about, 85, 136, 223, 241
   benchmarking, 255, 257–261
   business case, 244–245
   competitive analysis, 252–254
   content strategy, 254
   green hosting, 246–249, 259
   performance optimization, 260–261
   product methodology, 244–245
   researching, 251
   sprints, 254
   testing options with, 139
   user experience and, 249–250, 254, 260
   vision and goals, 242–243
   web design considerations, 249–257, 260
Ecological Economics Research Trends (Weidmann and Minx), 230
Ecovative Design, 22, 27
Edison Electric Institute (EEI), 71
education about sustainability, 272–274
EEI (Edison Electric Institute), 71
efficiency(ies)
   identifying, 12
   Jevons Paradox on, 28
   renewables versus, xxv–xxvi
   supply chain improvements, 85
EFs (emissions factors), 15
EIG (Endurance International Group), 102–103
electricity usage
   calculating emissions from, 18
   carbon footprint, 239–240
   data centers and, 38–39, 67
   data disposal and, 44
   digital business processes and, xxii–xxiv, xxxi–xxxii, 18, 25, 164
   estimating needs, 15
   fossil fuels and, 73–74, 79–80
   Google and, 69, 140
   green hosting and, 69–75, 78–80
   hardware and, 270
   identifying efficiencies in, 12
   Internet and, xviii, xxvii–xxviii, 3–4, 19, 284
   Internet of Things and, 29–32
   loss in transmission, 43, 235
   metrics for, 81
   online searches and, 140, 169
   open source solutions, 96
   video streaming and, 35, 125–127
   virtual reality and, 277
   web design and, 55, 152, 154
electronics industry, 23–24
Ellen MacArthur Foundation, 276
Elmieh, Baback, 24
embedded fonts, 172
emissions factors (EFs), 15
Endurance International Group (EIG), 102–103
Energy, Department of, xxv, 269
EPA (Environmental Protection Agency), 77, 234
EPt(GHG) certification, 15
ESG (environmental, social, governance) issues, 19
ETags (Entity Tags), 211
Etsy, xxix–xxx, 145, 264
e-waste
   Belgium and, 272
   defined, 26
   Ghana and, 9, 272
   United States and, 47
Expires header, 210–211
Exploit Scanner plug-in, 205
ExxonMobil, xv
Exygy, xxix, 301–302

F

Facebook
    data centers and, xxii, 38, 40, 145
    open source practices, 95–96
    renewable energy and, 40, 147
    site usage, 107
    WordPress plug-ins, 205
Fairphone, 9, 271–272, 296–297
Fast Company, 255
favicon.ico, 213
Federal Communications Commission, 38–39
Feldman, Jeffrey, 97
Fences module (Drupal), 208
file management, 85
findability of content, 59–60, 205, 248, 259
Flash Player, 188, 249–250
Flickr, 145
flush() function (PHP), 212
fonts (visual design), 171–172
Format community, 155
Formstack 508 Checker, 97, 223
fossil fuels
    Amazon Web Services and, 68
    Apple Computer and, 67
    carbon footprint example, xvi–xviii
    digital products and services, 78, 233
    divesting from, xvi
    electrical grid and, 73, 79
    "going green" campaigns, 20
    political environment and, 70
frameworks versus libraries, 198–200
Frick, Tom, 301
Frost, Brad, 227
Future Friendly, 53–54
FYA.tv, 300–301

G

GCP (Google Cloud Platform), 69
GDC, 272
Gelfand Partners Architects, 87
GeSI (Global e-Sustainability Initiative), xxii
GET request, 212
Ghana, e-waste and, 9, 272
GHGs (greenhouse gases)
    about, 230
    electricity and, 19
    environmental impact of, xiii
    EPt(GHG) certification, 15
    Greenhouse Gas Protocol, 18–19
    ICT and, xviii–xxi
    Internet and, 3–4
    LCAs calculating, 18–19
    measuring, xviii
    systems-thinking approach, 6
Gifford, Mike, 94–95, 225, 292–293
GIF format, 174
GitHub, 96, 219, 252
Global e-Sustainability Initiative (GeSI), xxii
global warming. *See* climate crisis
global warming potential (GWP), xviii
goals (LCAs)
    about, 13–14
    choosing quantification approach, 15
    considering certification, 15
    defining scope, 14
    engaging stakeholders, 15
    setting baseline, 15
    setting boundaries, 14
GoDaddy web hosting, xxiii
Goerlich, Kai, xxii
Goldberg, Michael, xxii
Goliath/Frank (Winter Storm), ix–xi
Google
    AMP project, 148, 205
    data centers and, xxii, 38, 40
    environmental impact website, 35
    performance optimization, 191, 213–214
    renewable energy and, xxii, 35, 40, 140, 147
    responsive design and, 187
Google+, 145
Google Analytics
    about, 44
    content audit example, 117–118
    content strategy and, 113, 131–132
    Greenanalytics and, 252
    inventory analysis and, 16
Google Cloud Platform (GCP), 69
Google Fonts, 172
Google Hangouts, 127
Google Hosted Libraries, 200

Google PageSpeed Insights, 200, 210, 213–214, 223, 247
GOOs (Guarantees of Origin). *See* RECs (renewable energy credits)
Gore, Al, 292
governance
  as business function, 10
  SASB sustainability standards for, 19
Green America, xxix, 72, 77, 314
Greenanalytics tool, 252
Green Boilerplate template, 93–94, 311
green building movement, 7–8
Green Code Lab, 252
The Green Grid, xxxii, 266
green hosting
  challenges of, 70–74
  customer service challenges, 101–104
  Ecograder case study, 246–249, 259
  electrical grids and, 72–74
  fonts and, 173
  future-friendly web and, 266–269
  good company versus good marketing, 79–81
  potential barriers and workarounds, 98–104
  RECs versus renewables, 74–77
  reliability challenges, 81, 100–101
  renewable energy and, 57–58, 65, 70–79, 98–99, 103–105, 128, 246–247
Green House Data, xxix, 75–76, 100, 297–298
greenhouse gases. *See* GHGs (greenhouse gases)
Greenhouse Gas Protocol, 18–19
Greenpeace International, 51
Greenpeace USA
  about, 309
  Click Clean Scorecard app, xx, xxix
  Clicking Clean annual report, xviii–xix, 35–36, 67, 71, 103–104, 145
  on data centers, 77
  on hardware sustainability, 271
  RECs versus renewables, 75–76
  social strategies, 146–147

green tariffs, 74
greenwashing, xxiii, 20–21, 64
Greenwashing Index, xxiii
Green Web Foundation, 246, 266, 268, 312
grist.org, 7
Grunwald, Michael, xxv
Guarantees of Origin (GOOs). *See* RECs (renewable energy credits)
GWP (global warming potential), xviii
Gzip tool, 192, 210

H

Halvorson, Kristina, 110
hardware
  disposal considerations, 47
  future-friendly Internet and, 270–272
  updating, 45–46
hardware disposal (virtual LCAs), 47
Haugen, John, 239–240, 313
Heap Media, 169
helix of sustainability, 10–11
Hemmings, Greg, 125, 128, 279, 300–301
Hemmings House Pictures, 125–127, 300–301
Heroku cloud provider, 40
Homekit framework (Apple), 31
Honeyman, Ryan, 308
Hootsuite, xxix, 145, 264
HostGator web hosting, xxiii
Howe, Neil, 310
HTML/HTML5 standards
  about, 42
  accessibility and, 183–184
  adoption of, 222
  AMP project, 147
  Ecograder tool and, 245
  HTTP requests and, 247
  inline images, 177
  performance optimization, 195
  performance optimization and, 205, 209–215
  responsive design and, 187–188
HTTP Archive, 2, 192, 235, 252
HTTP requests
  carbon footprints and, 247
  CSS sprites and, 176–177

INDEX | 323

Ecograder case study, 247
load speed and, 171
performance optimization and, 200, 203, 206–213, 218–221
Hubspot, 113, 242
Huffington Post, 35
Hulu
growth expectations, 108
site usage, 108
video streaming, xxi, 35
VR content, 37
Hurricane Sandy, 101
hydroelectric power, 40, 67, 73
Hynds, Davo, 204

I

ICA (International Co-operative Alliance), 88
ICROA (International Carbon Reduction and Offset Alliance), 294
ICT (information and communications technologies), xviii, 266
ImageCache module (Drupal), 208
Image Optimizer, 175, 260
imagery (visual design)
about, 173
CSS sprites, 176
format considerations, 174
image compression, 175, 204
inline images, 177–178
impact assessment (LCAs)
about, 12–13, 18
B Impact Assessment, xxix, xxxii, 22
informationalization strategy, 152
information and communications technologies (ICT), xviii, 266
information architecture, 118–120
infrastructure layer, 153
Inhofe, James, xv
inline images, 177–178, 210
innovation in business, 22–25
Instagram, 108, 145
interaction (virtual LCAs), 44–45
International Carbon Reduction and Offset Alliance (ICROA), 294

International Co-operative Alliance (ICA), 88
International Institute for Sustainable Development, 6
Internet Live Stats, 140
Internet of Things (IoT), 25, 29–34
Internet sustainability
about, 28
analytics and IoT, 275–277
conscientious companies, 264–266
data centers, 38–40
education and, 272–274
electricity usage and, xviii, xxvi–xxvii, 3–4, 19, 284
green hosting, 266–269
hardware considerations, 270–272
incubation and, 275
industry leaders predictions, 279–284
IoT, 29–34
Jevons Paradox, 28
online legacy, 278
runaway page growth, 34
video streaming, 34–36
virtual LCAs, 44
virtual reality and, 37–38, 277–278
interpreting data (LCAs), 12–13, 18
inventory analysis (LCAs), 12–13, 16–18, 43–44
InVision tool, 160
IoT (Internet of Things), 25, 29–34, 275–277
ISO 17024 certification, 15

J

Jacobs, Ian, 314
Janofski, Eric, 312
JavaScript
AMP project and, 147
Ecograder tool and, 245
performance optimization, 195, 199, 201, 204, 206, 208, 211–213, 220
WordPress API, 206
Jevons Paradox
about, 28
content strategy and, 110, 140

performance optimization and, 227
video streaming example, 36
Jevons, William Stanley, 28
JPG format, 174–175
jQuery library, 208

K

Katz, Josh, 51
Kauffman Index, 68
Kemp, Simon, 230–231
Kerry, John, 51
Keyword Explorer tool, 142
keywords in online searches, 141–143
Kickstarter, xxix, 264
Klein, Laura, 153
Klein, Naomi, 4, 293
Kraken.io compressor, 175
Kramer, Kem-Laurin, 152, 181, 314
KTH Royal Institute of Technology (Sweden), 252

L

Larsen, Todd, 72, 76–77, 314
Lawrence Berkeley National Laboratory, 234
Lazy Loader module (Drupal), 208
lazy load images, 204, 208
Lazy Load plug-in, 204
Lazy Load XT plug-in, 204
LCAs (life cycle assessments)
    about, 12–13, 231–232
    example using, 26–27
    Greenhouse Gas Protocol, 18–19
    impact assessment, xxix, xxxii, 12–13, 18, 22
    interpreting data, 12–13, 18
    inventory analysis, 12–13, 16–18, 43–44
    power of users, 15
    setting goals and scope, 13–15
    virtual, 41–48
LCIs (Life Cycle Inventories), 16–18
Lean Analytics (Croll and Yoskovitz), 136
lean personas, 158–159
Lean Startup movement, 92–93, 153, 221–222
libraries versus frameworks, 198–200

Lifecourse Associates, 310
life cycle assessments. *See* LCAs (life cycle assessments)
LimeRed Studio, 84–85, 91, 135, 298–299
LinkedIn, 36, 96, 145, 159
Linuxcon, 96
listicles, 137
Livefyre comments system, 217
Lonigro-Boylan, Emily
    about, 298–299
    on Agile methods, 91
    on content strategy, 135
    on environmentally-friendly workspaces, 84–85
    on future-friendly Internet, 279

M

MadPow.com, 311
maintenance versus performance, 197–198
Mall, Daniel, 164
Manoverboard, xxix, 79–80, 291–292
manufacturing industry, xxii–xxiv
manufacturing (LCA component), 17, 41
Marcotte, Ethan, 186
marketing
    as business function, 10
    good company versus, 79–81
    sustainability initiatives in business, 20–21
Marketing Grader tool, 242
Markiewicz, Pete
    about, 310–311
    on design process, 156, 196
    Green Boilerplate template, 93–94, 311
    on keywords, 141
    on misinformation, xxvii
    "Save the Planet Through Sustainable Web Design" article, 152, 241
    on sustainability education, 272–273
    on sustainable Internet, 281–282
    on user experience, 151–152, 254
    on virtual LCAs, 41, 44, 231–233
    on virtual reality, 38
    on WordPress JavaScript API, 206

materials (LCA component), 17, 22, 41
McDonough, William, 276
McGrane, Karen, 112
Memcached module (Drupal), 209
methane
    carbon footprint, 230–231
    GWP of, xviii
metrics (measurement)
    content audit, 114–118
    Ecograder case study, 245–246
    navigation considerations, 120
    performance optimization, 200–201
    selecting, 136
    tracking for content, 113–114
    user experience, 180
Meyer, Pamela, 130–131
microgrids, 79
Microsoft, 40
Mightybytes
    about, xxxi, 315, 87
    Climate Ride website and, 257–261
    design deliverables, 167–169, 172
    design workflows, 155
    Ecograder case study. *See* Ecograder
    environmentally-friendly workspaces, 83–84
    green hosting search, 98, 103
    mission-driven work, 21
    NACSA component library, 163
    performance optimization, 204, 215, 217
    stakeholder model and, 85
Mikkelson, Eric, 191, 194, 197, 215
Mills Office, 127
Mills, Shawn, 75–77, 237–238, 279, 297–298
Minify module (Drupal), 208
Minimum Viable Businesses, 275
Minimum Viable Product (MVP), 92, 243, 250
MinQueue plug-in, 204
Minx, Jan, 230
mission statements, 86–88
mobile devices
    AMP project, 147
    bandwidth considerations, 185
    carbon footprints, 229, 234, 237
    creating mobile pages, 205
    Green Boilerplate template, 93
    Internet usage and, 144–145
    low power mode, 166
    mobile-first strategy, 42, 184–187
    OLED screens, 170
    performance optimization, 34, 147, 213, 249–250
    progressive enhancement, 185
    proprietary technologies and, 188
    responsive design, 186–187
    search queries, 144, 187
    video streaming, 124–125
Month16, 275, 307
Moz metrics platform, 113, 142, 249
MozRank, 249
MVP (Minimum Viable Product), 92, 243, 250

N

Naidoo, Komi, 51
NASA (National Aeronautics and Space Administration), xiii
Nascent Objects, 23–26
Nav Flow Test tool, 120
Nest thermostat, 25, 30, 31–33
Netflix
    growth expectations, 108
    performance optimization, 192
    renewable energy and, 145
    site usage, 108
    video streaming, xxi, 34–35
    VR content, 37
network downloads (virtual LCAs), 43
NOAA (National Centers for Environmental Information), xi
Norton, Quinn, xxvi
nuclear power, 35, 40, 73, 268

O

Obama, Barack, xxv
OLED screens, 170
online searches
    on mobile devices, 144, 187
    sustainable, 139–146, 169
onsite search engine, 143–144
Open Concept Consulting, 94–95, 292–293
open source framework
    about, 94–95

addressing challenges with, 95–96
B Corporations and, 293
Fairphone and, 272
potential barriers and workarounds, 98
sustainable communities and, 95–96
virtual reality, 42
Open Web Group, 225
operations (business function), 10
Optimal Workshop, 120, 180
Optimizely, 180
Orbit Media, 111, 304–305
OSVR framework, 42
O'Toole, Greg, 314
Our Common Future paper, 4–5

P

P3 Plugin Performance Profiler, 203
PaaS (Platform-as-a-Service), 69
packaging (LCA component), 17, 22, 41
page bloat, 217–218
page briefs, 162–163
PageFrog, 205
Pandora, 145
Paris Agreement on Climate Change, 50–51
performance optimization
 accessibility and, 225–227
 CMS optimization, 201–209
 collaboration and, 196–197
 Ecograder case study, 260–261
 finding perfect balance, 191–193
 libraries versus frameworks, 198–200
 maintenance versus, 197–198
 mobile devices and, 34, 147, 213, 249–250
 performance rules, 210–214
 potential barriers and workarounds, 227–228
 runaway page growth and, 34
 speed as metric, 200–201
 VR content and, 38
 workflow tips, 221–222
 WPO, 194–195
Petricek, Tomas, 198
PGE (Portland General Electric), 71
Photoshop tool, 175

PicMonkey tool, 260
Pingdom Tools, 200, 223
Pinterest, 35, 145, 147
planning versus preparing, 133–135
Platform-as-a-Service (PaaS), 69
PNG format, 174
Pollack, Jill, 121–122, 306
polystyrene, 22
Pomerantz, David, 75–77, 271, 280, 309
Porter, Eduardo, xxii
Porter Ranch methane leak, xviii
Portland General Electric (PGE), 71
Post, René, 268–269, 312
POST request, 212
Power API, 252
power consumption. *See* electricity usage
power purchase agreements (PPAs), 74
Power Usage Effectiveness (PUE), 267
PPAs (power purchase agreements), 74
preparing over planning, 133–135
presentation layer, 153
printers and printing
 3D printers, 23–24
 print styles, 178–179
process (business function), 10
Product Science, 30, 88, 240–241, 313
progressive enhancement, 42, 185
proto-personas, 158–159
Public Benefit Corporations, 103–104
PUE (Power Usage Effectiveness), 267

R

Rao, Srinivas, 131
raster images, 174
RECs (renewable energy credits)
 B Corporations and, 265
 defined, 74
 green hosting and, 246, 267
 renewable energy versus, 75–78, 105
Reddit, 35, 102, 145
reliability
 in green hosting, 81, 100–101

INDEX | 327

performance optimization and, 214–221
Renewable Choice, 77
renewable energy
   Amazon Web Services and, 65, 105, 164
   Click Clean Scorecard, xx, xxix
   data centers and, xxii, 30, 40, 67–69, 75–79, 105
   efficiency versus, xxv–xxvi
   Facebook and, 40
   future-friendly web and, 263–269, 276–277
   Google and, xxii, 35, 40, 140
   green hosting and, 57–58, 65, 70–79, 98–99, 103–105, 128, 246–247
   mindset shift toward, 54–55
   Paris Agreement on Climate Change, 50–51
   powering the Internet with, xxii
   RECs and, 74–77, 105
   social media strategies and, 145–147
   Twitter and, xx
   video content and, 124
repurposing content, 137–138
Requests for Proposals (RFPs), 89
responsive design, 186–187
retail and consumer products industry, xxii–xxiv
revision control, 204–205
Revision Control plug-in, 204–205
RFPs (Requests for Proposals), 89
Ritzrau, Will, xxii
Roach Motel, 165
Rochdale Principles of Cooperation, 88
Ruby on Rails framework, 92
Rupert, Dave, 200–201

S

SaaS (Software as a Service), 147
Salva, Miquel Ballester, 279, 296–297
SASB (Sustainability Accounting Standards Board), 19
"Save the Planet Through Sustainable Web Design" (Markiewicz), 152, 241
scope (LCAs)
   about, 13–14
   choosing quantification approach, 15
   considering certification, 15
   defining, 14
   engaging stakeholders, 15
   setting baseline, 15
   setting boundaries, 14
Scorecard app (Click Clean), xx, xxix
search engine marketing (SEM), 140–141
searches
   on mobile devices, 144, 187
   sustainable, 139–146, 169
Secure Sockets Layer (SSL) certificates, 219, 221
SEM (search engine marketing), 140–141
SEOChat, 207
SEO Checklist module (Drupal), 208
SEO (search engine optimization)
   data centers and, 141
   Drupal module, 208
   Ecograder case study, 255
   inline images and, 177
   standards for, 205, 208
   storytelling and, 122
   sustainability and, 141–143
   web design considerations, 59–60, 63–65
   WordPress plug-in, 205
server uploads (virtual LCAs), 43
sharding, 221
shared resources, bandwidth considerations, 248
Shedroff, Nathan, 6, 54, 151–152, 230
Skype, 127
Smush.it tool, 175, 260
Snapchat, 108, 145, 167
social media sites. *See* specific sites
software and visual assets (virtual LCAs), 42–43, 45–46
Software as a Service (SaaS), 147
software frameworks, 92–93
solar energy, 67–68, 71–74, 79, 104
Souders, Steve, 195
SoundCloud, 35, 145, 239
speed (performance optimization), 213–221
Speedy module (Drupal), 208
Spool, Jared, 110

sprints (Agile), 91, 221–222, 250–251, 254
Squarespace, 101
SRI (Stanford Research Institute), 86
SSL (Secure Sockets Layer) certificates, 219, 221
stakeholders
  as business function, 10
  engaging, 15
  model overview, 85–86
standards
  accessibility, 42, 97, 183–184
  appliance and equipment, xxv
  energy consumption, 128
  Greenhouse Gas Protocol, 18–19
  IoT, 30
  SASB, 19
  SEO, 205, 208
  web, 52–55, 97–98, 183–188, 222–223
Stanford Research Institute (SRI), 86
Stevens, Robert, 76–77, 294
Stockholm Syndrome, 167
StoryStudio Chicago, 121, 306
storytelling in content strategy, 120–123, 164
Strauss, William, 310
style tiles (visual design), 168–169
Sucuri Security plug-in, 204
Sullivan, Timothy Michael, 240
supply chains
  B Impact Assessment and, xxx
  challenge of, 269
  environmental disasters and, 270
  greening up, 112–113
  inventory analysis and, 16
  reevaluating, 85
SurveyMonkey, 136
sustainability
  benchmarking, 19
  building sustainable solutions, 1–3, 9
  business. *See* business sustainability
  carbon footprints. *See* carbon footprints
  content strategy. *See* content strategy
  defined, 4–6
  environmentally-friendly workspaces, 82–85
  green hosting. *See* green hosting

  hardware disposal and, 47
  Internet and. *See* Internet sustainability
  Lean/Agile workflows, 88–92
  mission statements, 86–88
  open source and, 94–97
  performance optimization and. *See* performance optimization
  software frameworks, 92–93
  stakeholder model, 85–86
  systems-thinking approach, 5–8
  user experience and. *See* UX (user experience)
  virtual LCAs. *See* virtual LCAs
  web design. *See* web design and development
  web standards, 52–55, 97–98, 183–188, 222–223
Sustainability Accounting Standards Board (SASB), 19
Sustainable Brands, 255
Sustainable St. John, 127
Sustainable UX conference, 311
Sustainable Web Ecosystem Design (O'Toole), 314
SVG format, 174
system fonts, 171
systems-thinking approach, 5–8

T

task layer, 153
Technology Innovation Management Review, 95
TedX Talk, 255
Tesla Motors, 96
Third Partners, LLC, 239–240, 313
This Changes Everything (Klein), 4
3D printers, 23–24
Thrive Design, 7–8
transmaterialization strategy, 151
transportation and logistics industry, xxii–xxiv
Treejack tool, 120
tree testing, 113
Trello tool, 164
Triple Pundit, 145, 295
Tufekci, Zeynep, 45
Tumblr, 145
Tweetfarts, xx
Twitter, xx, 145, 147–148

Typekit font service, 172–173
typography (visual design), 171–172

U

UIE (User Interface Engineering), 110
UNESCO, 266
United Nations
    Conference on Climate Change, 49
    Information Economy Report, xxi
    Paris Agreement on Climate Change, 50–51
updating software and hardware, 45–46
Uptake (startup), 275
Usability Hub, 120, 180
usage (LCA component), 17, 41
US Energy Information Administration, xxviii, 40, 43
user experience. *See* UX (user experience)
User Experience in the Age of Sustainability (Kramer), 152, 181, 314
User Interface Engineering (UIE), 110
UserZoom tool, 120
utilities industry, xxii–xxiv
Utz, Emily, 303–304
UX for Lean Startups (Klein), 153
UX (user experience)
    about, 153–155
    accessibility considerations, 181–182
    Agile practices and, 157
    avoiding dark patterns, 165
    component design, 163
    content patterns, 162–163
    content strategy, 112, 162–167
    defining users and devices, 158–159
    display patterns, 163
    Ecograder case study, 249–250, 254, 260
    measuring success, 180
    page briefs, 162–163
    potential barriers to, 188
    print styles, 178–179
    sustainable design workflows, 155–156
    user stories, 164
    users versus life cycles, 151–153
    virtual LCAs, 44–45
    visual design. *See* visual design
    VR content and, 38
    web design considerations, 60–61, 63–65, 155–156, 158–167
    wireframing tools, 160–161

V

validating work, 223–224
Varnish module, 209
Vasquez, Amber, 155, 158, 160, 167
vector images, 174
version control, 214–215
video content
    accessibility of, 129
    popular platforms for, 108
    story diet, 124–125
    strategies for, 124
    streaming, 34–36
    video compression, 128–129
    video workflows, 125–128
Vimeo, 35, 128, 145
Vine, 35, 108, 145
virtualization, xxi–xxiv, xxxii
virtual LCAs (VLCAs)
    about, 41–42, 231–233
    design and development, 42
    disposal of data, 44–46
    hardware disposal, 47
    interaction, 44–45
    network downloads, 43
    server uploads, 43
    software and visual assets, 42–43
virtual reality (VR), 37–38, 42, 277–278
visual design
    about, 167–168
    color choices, 169–170
    fonts and typography, 171–172
    imagery, 173–177
    style tiles, 168–169
VLCAs. *See* virtual LCAs (VLCAs)
Volkswagen, 20
VR (virtual reality), 37–38, 42
Vudu, 35

W

W3C (World Wide Web Consortium)

about, 52
Markup Validation Service, 97, 205, 223
on video accessibility, 129
Web Content Accessibility Guidelines, 181–182
web standards and, 183–184, 222
W3 Total Cache plug-in, 204
Waag Society, 271
Waldern, Jonathan, 37
waterfall workflows, 88–90
WCAG (Web Content Accessibility Guidelines), 181–182
The Weather Channel, ix–xi
weather versus climate, xiii–xiv
Web Content Accessibility Guidelines (WCAG), 181–182
web design and development. *See also* content strategy
about, 49–52
average page size, 2
collaboration in, 112, 155–156, 196–197
creating mobile pages, 205
Ecograder case study, 249–257, 260
electricity usage and, 55, 152, 154
greenwashing and, 20–21
keywords and phrases, 142–143
leading by example, 123–124
performance optimization and, 222–223
potential barriers and workarounds, 63–65
runaway page growth, 34
standards for, 52–55, 97–98, 183–188, 222–223
suggested framework, 55–63
user experience and, 60–61, 63–65, 155–156, 158–167
in virtual LCAs, 42
Web Energy Archive Ranker, 252
web fonts, 171–172
Webpagetest, 252
web performance optimization (WPO), 61–65, 194–195
WebVR framework, 42
Weidmann, Thomas, 230
Williams, Ian, 230–231
Wilson, Fred, 194
wind energy, 71, 76
wireframes, 160–161, 167, 254

Wissner-Gross, Alex, 240
Wordfence plug-in, 204
WordPress
about, 203–207
CMS usage, 201–203
JavaScript API, 206
search support, 143
workflows
Agile methodology, 90–91, 221–222
cone of uncertainty and, 89
Lean Startup movement, 92–93, 221–222
performance optimization and, 221–224
sustainable web design, 155–156
video, 125–128
waterfall, 88–90
workspaces, environmentally-friendly, 82–85
World Economic Forum, 276
World Resources Institute, 18–19
World Wide Web Consortium. *See* W3C (World Wide Web Consortium)
World Wide Web Foundation, 266
WP DB Manager plug-in, 204
WP-Optimize plug-in, 204
WPO (web performance optimization), 61–65, 194–195
WP Smushit plug-in, 204
Wright, Laurence A., 230–231
Wroblewski, Luke, 184
W.S. Badger Company, 87
WyoREC, 77

X

XMLHttpRequest, 212
XML standard, 183

Y

Yahoo!, 148
Yelp, 35
Yoast Optimize DB plug-in, 204
Yoast SEO plug-in, 205
Yoskovitz, Benjamin, 136
YouTube
hosting services, 128, 239
Internet traffic and, xxi
site usage, 108
video streaming, 34–35, 108, 129

Yu, Bernard, 155, 270

Z
Zimmerman, J.B., 166–167
Zipcar, 96

[About the Author]

Tim Frick is CEO of Mightybytes, a certified Illinois B Corp that builds creative digital solutions for conscientious companies. Mightybytes is committed to solving social and environmental problems with its work and is a leader in sustainable web design.

O'Reilly Media, Inc.介绍

O'Reilly Media通过图书、杂志、在线服务、调查研究和会议等方式传播创新知识。自1978年开始，O'Reilly一直都是前沿发展的见证者和推动者。超级极客们正在开创着未来，而我们关注真正重要的技术趋势——通过放大那些"细微的信号"来刺激社会对新科技的应用。作为技术社区中活跃的参与者，O'Reilly的发展充满了对创新的倡导、创造和发扬光大。

O'Reilly为软件开发人员带来革命性的"动物书"；创建第一个商业网站（GNN）；组织了影响深远的开放源代码峰会，以至于开源软件运动以此命名；创立了Make杂志，从而成为DIY革命的主要先锋；一如既往地通过多种形式缔结信息与人的纽带。O'Reilly的会议和峰会集聚了众多超级极客和高瞻远瞩的商业领袖，共同描绘出开创新产业的革命性思想。作为技术人士获取信息的选择，O'Reilly现在还将先锋专家的知识传递给普通的计算机用户。无论是书籍出版、在线服务还是面授课程，每一项O'Reilly的产品都反映了公司不可动摇的理念——信息是激发创新的力量。

业界评论

"O'Reilly Radar博客有口皆碑。"
　　　　——Wired

"O'Reilly凭借一系列（真希望当初我也想到了）非凡想法建立了数百万美元的业务。"
　　　　——Business 2.0

"O'Reilly Conference是聚集关键思想领袖的绝对典范。"
　　　　——CRN

"一本O'Reilly的书就代表一个有用、有前途、需要学习的主题。"
　　　　——Irish Times

"Tim是位特立独行的商人，他不光放眼于最长远、最广阔的视野，并且切实地按照Yogi Berra的建议去做了：'如果你在路上遇到岔路口，走小路（岔路）。'回顾过去，Tim似乎每一次都选择了小路，而且有几次都是一闪即逝的机会，尽管大路也不错。"
　　　　——Linux Journal

出版说明

随着计算机技术的成熟和广泛应用,人类正在步入一个技术迅猛发展的新时期。计算机技术的发展给人们的工业生产、商业活动和日常生活都带来了巨大的影响。然而,计算机领域的技术更新速度之快也是众所周知的,为了帮助国内技术人员在第一时间了解国外最新的技术,东南大学出版社和美国O'Reilly Meida, Inc. 达成协议,将陆续引进该公司的代表前沿技术或者在某专项领域享有盛名的著作,以影印版或者简体中文版的形式呈献给读者。其中,影印版书籍力求与国外图书"同步"出版,并且"原汁原味"展现给读者。

我们真诚地希望,所引进的书籍能对国内相关行业的技术人员、科研机构的研究人员和高校师生的学生和工作有所帮助,对国内计算机技术的发展有所促进。也衷心期望读者提出宝贵的意见和建议。

最新出版的影印版图书,包括:

- 《学习 OpenCV 3(影印版)》
- 《数据科学:R 语言实现(影印版)》
- 《数据驱动设计(影印版)》
- 《设计数据密集型应用(影印版)》
- 《Scikit-Learn 与 TensorFlow 机器学习实用指南(影印版)》
- 《Perl 语言入门 第 7 版(影印版)》
- 《Python 漫游指南(影印版)》
- 《算法技术手册 第 2 版(影印版)》
- 《RxJava 反应式编程(影印版)》
- 《设计人见人爱的产品(影印版)》
- 《你好,创业公司(影印版)》
- 《可持续性设计(影印版)》
- 《基础设施即代码(影印版)》
- 《Cassandra 权威指南 第 2 版(影印版)》
- 《网站运维工程(影印版)》
- 《商业数据科学(影印版)》
- 《深度学习(影印版)》
- 《深度学习基础(影印版)》
- 《面向数据科学家的实用统计学(影印版)》
- 《高性能 Spark(影印版)》
- 《Python 数据分析 第 2 版(影印版)》
- 《Unity 移动游戏开发(影印版)》